Women Managers:
Travellers in a Male World

Women Managers:
Travellers in a Male World

Judi Marshall
Lecturer in Organizational Behaviour,
School of Management,
University of Bath

JOHN WILEY & SONS
Chichester · New York · Brisbane · Toronto · Singapore

Library of Congress Cataloging in Publication Data

Marshall, Judi.
  Women managers.
  Includes index.
1. Women executives.   I. Title.
HF5500.2.M248 1984   658.4'09'088042   83–23579

ISBN 0 471 90419 8

British Library Cataloguing in Publication Data

Marshall, Judi
  Women managers.
  1. Women executives
  I. Title
  658.4'2'088042   HF5500.2

  ISBN 0 471 90419 8

Phototypesetting by Inforum Ltd, Portsmouth
and printed at The Pitman Press, Bath, Avon.

Contents

Contents

Acknowledgements

My thanks are due to the many people who have contributed in diverse ways to nurturing and stimulating the growth of this book. I wish especially to thank Peter, Paul, Angie, Howie, Gwenda, Richard, and the managers who talked to me about their lives.

I should also like to thank the Equal Opportunities Commission for funding the research study around which this book is written. The views expressed here are mine, and do not necessarily represent the views of the Equal Opportunities Commission.

Chapter 1

Introducing my Topic, and Myself

The backcloth to my writing is fertile chaos. This book emerges out of conflicts and contradictions. These appear both in the public arena, where the place of women in society is now much debated, and in the personal lives of many women, whether or not they are active in any of the Women's Movement's various manifestations. Here I shall be drawing on both these levels of analysis to consider the place of women in management. As these topics are so broad, complex and continually changing, my aim is to identify key issues, and explore the options, rather than come to 'conclusions'. I am not, then, advocating particular strategies or objectives. What is appropriate is an individual matter. I do, of course, have a sense of what is appropriate for me, now. I shall share my own current opinions — and continuing questions and uncertainties — with the reader, but they are offered simply as such.

It will soon become apparent that this is most definitely not a 'How to' (make it in management) book. It is rather a 'What do you want?' and 'Who are you?' book. In it I am addressing individuals initially, but organizations and society a close second. In this introductory chapter, I shall first give a flavour of the fertile chaos which forms the book's setting. Some appreciation of the context is essential to a satisfactory exploration of focal issues. There are two aspects of its setting which particularly explain why I am writing this book. One strand is the 'public' issue of the movement of women into management jobs; the other is the relevance to me personally of looking at women's issues. Later, I shall introduce the book's main themes, and outline the choices I faced in writing.

STRAND 1: RECENT TRENDS RELATED TO WOMEN'S ACCESS TO MANAGEMENT JOBS

There is currently genuine puzzlement and deep-seated confusion amongst commentators about what is happening to women's place in employment. This specific area is also interpreted as a barometer for their place in society generally. Officially equal opportunities for women are promoted through government legislation and publicity, and private companies' policies. But

1

2

these efforts are contradicted by their failure to have the type of effects people expected. The revised Women's Movement has been active in its many new guises in Britain since roughly 1968, and the Equal Pay and Sex Discrimination Acts have been on the statute books since 1970 and 1975 respectively. Yet if we look to key areas in which they might have had an impact, little seems to have changed in that time. On pay, the right to employment, and opportunities for top jobs, as prime examples, there are reports that women are actually losing ground in relation to men. In a review article of developments in Britain, Gibb (1981) calls the last five years one of the 'bleakest' periods in women's history. Available evidence suggests their fortunes have gone backwards.

Pay, and the Right to Employment

Having previously crept up towards those of men, women's earnings have dropped slightly and levelled off at around 75 per cent of those of men. Women have suffered a disproportionate share of unemployment in our recent wave of recession. More than 800,000 women lost their jobs between 1979, when a peak of 9.3 million were in work, and 1982 (Senker, 1982). They are now 29 per cent of the registered unemployed, compared with 22 per cent five years ago, and this does not include an estimated one million unregistered women.

Women and top jobs

Despite official policies, the established pattern that women do relatively unskilled jobs at the bottom of organizations has not significantly shifted. Ninety per cent of working women are in low paid, low skill jobs (Senker, 1982). Women's capture of management jobs has also suffered recently. In 1981 one in twelve managers were women, compared with one in ten in 1975. Estimated proportions of women managers vary according to the base on which calculations are made. All estimates, however, show similar trends — either no increase or a decline. The above figures are at the upper end of the estimated range. Focusing on senior positions, the proportion of women in both the United Kingdom and the United States is generally agreed to be less than one per cent (Powell, 1980).

A further disturbing trend is that women are losing ground in those few occupations in which they had previously dominated management levels. Personnel and teaching are two of the professions which started out as female specialisms, but are now seeing an increasing proportion of senior jobs go to men. This trend is associated with the professions' rising status. Novarra (1980) reports that the Institute of Personnel Management had 29 women and 5 men as its founder members in 1913. By 1927 membership was 420, fewer than twenty of them men. By 1972, women comprised only nineteen per cent of Institute membership, and were concentrated in junior posts. Of a study sample of 90 Directors of Personnel, not one was female. Similar developments are happening in teaching. Moves towards more mixed-sex and larger schools are accompanied by a decline in the proportion of female head teachers.

Gibb (1981) reflects an appropriate confusion in how to respond to this disappointing picture. It is easy to identify weaknesses in the relevant legislation, and the Equal Opportunities Commission claims that no further progress is possible unless its provisions are strengthened (Equal Opportunities Commission's Sixth Annual Report, 1981, HMSO). Legislation is, however, only part of the total picture. Whatever its 'bite', legislation can only go so far. Dissatisfaction with the United States' much tougher legal base bears this out. It leaves indirect discrimination (practices which inherently put women at a disadvantage) and prejudiced attitudes untouched. These seem relatively entrenched, as material from other arenas illustrates. Particularly relevant to women's potential for management are their still impoverished experiences in formal education.

Sexual inequality in education

Reviewing the teaching material used in schools, writers conclude that girls and women are still grossly under-represented, and that where they do appear they are portrayed mainly in passive, domestic and home based roles. In an analysis of 225 reading scheme stories, Lobban (1976) found that the most active and strenuous thing girls did was to hop and skip. Whilst more options are nominally open to them, girls in mixed schools (which are currently on the increase) make more traditional subject choices, than do those in single-sex schools (Sarah *et al.*, 1980). Concerned careers counsellors continually express disappointment at the lack of ambition and creativity in girl school leavers' expectations.

Some writers applaud the increasing female enrolment on management courses at universities as a sign of change, and warn organizations that they must prepare to deal with the challenge this influx will represent. The proportion of women management students increased from 12 per cent in 1973 to 27 per cent in 1977 (Cooper and Lewis, 1979). Whilst encouraging, on closer examination even this is less of a revolution than it appears. When asked (532 women in 22 univerisities took part in Cooper and Lewis's questionnaire survey), a high proportion (43 per cent) reported that they felt at a significant disadvantage as women, and anticipated difficulties in finding satisfactory lives in organizations. Many seemed also to be making the 'traditional' female job choices of personnel (21 per cent) or specialist/expert careers. (Thirty-seven per cent favoured finance/ accounting, for example.)

In 1982, 37 per cent of undergraduates and 29 per cent of postgraduates at British universities were women. When they leave many enter the professions. Then they disappear. Along with many other concerned commentators, I should like to know whether we can hide them much longer. Looking further ahead in the management education system confirms women's lack of substantial impact. In two of the United Kingdom's leading post-experience management training establishments, the proportion of women students is less than four per cent, and has changed very little over the past decade (personal communication).

Review of Strand 1, and its implications here

The above are only a sample of the contexts in which women's progress towards equality fails glaringly to live up to its apparent promise. Neither official statute, nor a growing popular awareness of 'women's issues', seems to have been sufficient to produce radical change.

In writing this book I therefore set out to explore the world of women managers, appreciating that the issues I would encounter would by no means be clear-cut. I soon found I could not view 'women in management' as a narrow, bounded area. I needed continually to turn to a wider context and import alternative, complementary perspectives into the discussion. I have attached considerable importance throughout the book to understanding the appropriate context or 'ecology' of a particular issue or topic. This wider context took several forms. One was the general issue, such as society's values about work, or the nature of life within large organizations, raised by a focused examination of women in management jobs. Secondly, I found I needed to explore other aspects of women's lives to understand the qualities, motivations and priorities they bring to employment. This need was instructive — I had to appreciate women's lives as wholes. They would not meaningfully break down into constituent parts, as women's various roles were too closely intertwined. A third context — women's place in society — is never far away during the book's explorations. Women bring their femaleness, with its connotations and status in society, with them when they enter organizations.

One further aspect of the wider context which warrants early acknowledgement is the backcloth of recession, unemployment and technological change with which we now live. These may have disproportionately negative effects on women's lives. Concepts of 'equal opportunities' at work were relatively tolerated in the 1960s and early 1970s, when Western economies had slack resources, and faced labour shortages in certain areas. Changed circumstances mean that much of the apparent interest in equality has evaporated, and even women's right to work is now sometimes contested. Concern to hold on to whatever jobs they now have may deter women not only from demanding additional provisions (such as child-care facilities which would make working more practicable), but also from enforcing their current statutory rights. Fisher (1978) provides evidence of women's needs being assigned a low priority in this way. As the result of a job analysis exercise on 2000 jobs in the Health Services Centre, Winnipeg, Canada, pay increases of up to 50 per cent for women employees were deemed necessary. The Canadian Anti-Inflation Board quashed this recommendation, and allowed rectification only at a minimum level in line with strict 'equal work' assessments.

I find such challenges particularly threatening for the underlying values they reveal. Some sympathetic writers suggest that women can expect little attention now, but should wait, planning how to open opportunities for themselves once the economic situation becomes less critical. But this is an unequal way forward, assigning women's rights a lower priority than those of men. It can be interpreted as a reassertion of male dominance under a psuedo-rational guise aimed at defusing the Women's Movement's current energy. An alternative interpretation, which reaches similarly negative conclusions, is that society is facing a new problem but is failing to cope adaptively. Not only does it not recognize that problems are potential opportunities, but it is also attacking with outmoded and inappropriate coping strategies. As this book touches on this context too, in its broadest possible interpretation it is also about possible futures for society as a whole.

STRAND 2: THE PERSONAL RELEVANCE OF WRITING ABOUT WOMEN'S ISSUES

The second strand to this book's conception is the relevance to me personally of working in an area of women's studies. All writing is autobiographical in some sense, and here especially I want to acknowledge this, and draw on my personal understanding, and its recent developments, as a powerful sense-making process. The changes I have experienced have been influenced and accelerated, but were not initiated, by working professionally on the theme of women in management. Dealing with the topic was in fact largely accidental— if anything can be. I had previously avoided involvement in overtly 'women's issues', as I did not want to risk the stigma which seemed inevitably to accompany them. I have recently had ample evidence that others share this reluctance. Many people, particularly women, tell me this book is a dangerous venture as far as my academic career is concerned.

Since starting what I shall here call my 'explorations' or 'journey', life has been both intellectually and emotionally challenging and stimulating. I have tested all new material and ideas not only against the accounts of a research sample (introduced below), but also against my own views of the world. This was at times a personally disturbing experience — as my later comments on different approaches to women managers will show — but a uniquely enlightening one.

Development of my intellectual understanding has, then, been paralleled by the emergence of a new sense of myself as a woman in a public world structured largely by men. As that sounds a little dramatized, let me explain. This is a 'fact' I had partially ignored, and certainly played down previously— that seeming to be the most comfortable strategy for all concerned, including myself. I do not want you to think I thought I was a man. Rather I submerged any awareness of being a woman, especially at work, and generally hoped to be treated as a person first. I was often surprised when I discovered that my sex had been significant in some way to other people. I experienced a sense of dissatisfaction, for example, when after a talk on job stress to an outside group, the organizer's main thanks was that I had certainly been the most attractive speaker they had

had that year. I wondered what people had heard of the ideas which had been *my* main concern.

Somehow, though, I did not add up these occasional incidents to anything much, since I had no frameworks in which they really made sense. I found, however, that I experienced increasing pressure of this sort as I started to operate in a more public setting than previously. At the same time, I was increasingly aware of popular coverage of women's issues. It became time to start developing some more appropriate ways of making sense of my experience, and new strategies for being effective in difficult situations. 'Research', as it so often does, proved a useful vehicle through which to explore issues of key personal relevance.

Two further factors help explain why this exploration comes now in my life. I work in an environment, and with colleagues, that make personal exploration and development a relatively low-risk process. I am privileged in this. I have had the freedom to experiment with who I am, and the space and support with which to do it. Secondly, I am now on the foothills of a relatively predictable career path — if I want to follow it. That raises questions which will be familiar to many readers: Is it really what I want to do? Which aspects should I concentrate on? How committed do I want to be? What do I want to achieve in other areas of my life? Answering these satisfactorily requires reassessing who I am, arriving at a fresh sense of self.

My journey — and some of the places I visited

A few remarks about the nature of this journey will help explain its contribution to the book. In essence, I have spent the past few years in an often turbulent dialogue with the issue of women in society. The outside manifestations of this journey are simple enough. I was sponsored by the Equal Opportunities Commission to do a research study of women in management jobs. I did some reading on women and work and management in preparation, and eventually interviewed 30 women in two industries. A full introduction to the study is found in Chapter 5. I prepared a detailed report on my findings for the Commission, but then laid it aside for a while to attend to other matters. Later I dusted it off, negotiated a contract to write this book, and started reading again — this time more widely than before. It is the importation of the more radical perspective on women in society I then encountered, together with a need to question what others take for granted, which are the key things I have to offer here. Certainly it was at this point that the pace of the journey increased, as it began to make sense to my experience as well as my intellect.

I had started out with little more than a dulled dissatisfaction with the world as it is, and with no burning interest in women's issues. In this new phase of enquiry, I moved through phases of immersion in particular viewpoints, developing my own ideas and values as I went. As I played host to a perspective I would see and react to the world with its attitudes and values. The new-to-me various feminist perspectives had most impact on me in these terms. They plunged me into interpreting the world about me in radically new ways. Typically a given view's usefulness would disintegrate in time, as it became too straightforward to match the world as I found it. As a model broke down I would see more clearly which vestiges I wanted to retain, and would move on to experience another framework. Some ideas would be instantly appealing, feel so right. Others I would reject initially, coming gradually to a belief in a version I could call my own. My journey has therefore been sporadic and disconnected. There have been times of refinement, of clear, widely based knowing — particularly lately. This knowing has, however, been balanced and fuelled by movements into chaos, when I was unable to make sense of much, and felt swamped and overloaded. One of the many excitements has been the discovery that work I have done in other areas (such as job stress) is relevant, and sheds fresh light on, the new material I have encountered.

I should like to illustrate these processes of enquiry with a brief, personalized, introduction to the main streams of writing about women which have influenced this book. As far as women in management is concerned, two contrasting patterns can be identified in the frameworks through which analysis is conducted and action approached. In one, the 'reform' feminist viewpoint, the male world of activities and characteristics is tacitly used as a positive model to which women should aspire. Its values are not significantly challenged. Suggestions emerging from this literature are typically couched in terms of procedures which will allow women greater access to work, and to (senior) management positions. Women's suitability is a critical issue — hence the literature reviewed in Chapter 2, which seeks to demonstrate that women are closely similar to men and epitomizes this approach.

The writing in this vein is detached, cool, determined and practical; but sometimes idealized and often boring. I seldom found this approach truly absorbing. It seemed superficial, ignoring too many of the less-nice aspects of organizational life (like stress), of which I am highly aware from my own research and experience. At first I was surprised by the narrowness of the reform perspective; it now disappoints me. Even so, I found its values so insidious that they would temporarily creep into my own thinking— requiring me to think myself free once I realized. Using men as the standard by which to judge women's behaviour, and career advancement as a general-purpose measure of success, are the approach's most powerful assumptions. Having waded through volume after volume permeated with concern about 'how to make it', I would often feel quite battered, holding on to a dulled knowing that this was only a small chunk of life, but temporarily unable to see far beyond it.

The second main stream of literature, that of so-called 'radical' feminists, is wholly different in content and approach, and I found it more compelling. Its perspectives were new to me, and made sense of parts of my experience which had previously been without sense. This new literature made rich and enjoyable reading. The concepts and ideas were clear and competent, but they had not lost touch with the experience from which they emerged. This, too, was honoured and expressed. The radical perspective rejected using men as a model for women. It sought instead to understand women's inequality, their continuing low status in relation to men, and to reaffirm women's own sense of being. It depicted women as oppressed; with men, as the dominant force in society and as individuals who benefit from women's position, as their oppressors. One of the biases I temporarily adopted from these writers was to see men as all the same, conforming to the stereotype of dominating males. Everywhere I looked women were at an inherent disadvantage, put down. This made personal relationships difficult for a time. I was highly sensitive to being patronized, ignored, taken advantage of, criticized, mistaken for someone's assistant, or whatever. Looking with these new eyes, I found ample evidence to suggest that had I been male I would have been treated differently, better, in many situations. I tended to avoid male company during this stage, from indifference. Men's conversation did not have the lustre and importance I had once attributed to it.

I eventually found this perspective on the world too simple in its entirety as well. People became complex individuals again, and I moved on to explore anew. One outcome of this period I value particularly is a greater appreciation of women as people than I had experienced since adolescence. By temporarily rejecting men as desired company, and revising my valuation of society accordingly, it seems I have rediscovered 50 per cent of the population in whom I had recently shown only a limited interest.

In the pages which follow, I have borrowed selectively from these two streams. I find it impossible to agree wholeheartedly with either, or to reject either in its entirety. Rather, in making sense for myself, I have culled from both. This may leave me with apparent conflicts in my views. I do not consider this disastrous. The more heinous crime would be to neaten them up spuriously for public consumption. Neither has steering an apparently 'middle' course felt like compromise. Rather it has required exploring a wider frame of reference in which they could both be incorporated. In *The Aquarian Conspiracy*, Ferguson (1981) identifies an approach she calls 'radical centre'. This involves retaining the valuable elements in *both* of two apparently opposing positions — such as political parties — and viewing them dialectically, so that they can co-exist, be present in each other, and be reconciled at a higher level of explanation. In retrospect this aptly, if grandly, describes the activity I have been engaged in.

Reflecting on this process I find a third women's voice in and around me, seeking integration in this mêlée — perhaps seeking to calm some of the clamourings. It is the quiet, scarcely verbal, voice of what women have been and are. It asks us to remember and value the strengths they have developed and lived, and their ability, too, to be silent. It is a 'voice' we must not forget.

I offer this personal strand as a basic core to this work because I have found it a powerful sense-making apparatus. I suspect that it may also reflect the experiences of many other women who are partly involved in the (largely male) world of work; bombarded by different messages about the places for women; aware, if only vaguely, of sometimes being at a disadvantage; surrounded by a confusing array of different models of how to be a woman, and who are struggling to make personal sense of it all, *and* remain viable as people in their day-to-day work and home relationships. This 'strand' is an ongoing process of intellectual and personal development, not a set of opinions I have arrived at and expect to maintain 'forever'. Pausing in this process to write, I shall speak from my frameworks in their current states, whether clear and developed, or still riddled with questions as many are. Only by giving myself this freedom can I address adequately the complex issues which characterize this area of enquiry.

PUTTING MY TWO STRANDS OF ENQUIRY INTO PRACTICE

As a prelude to describing the book's layout, the next three sections outline key choices I faced in writing: who to speak to, how far into the past to delve, and how closely to focus on women alone.

My audience

My main audience in this endeavour, as I have already intimated, is people like myself — women who, whatever their age and apparent circumstances, are assessing or reassessing their position in relation to a world of employment which has so far been dominated by men's needs and values. It is not about or for a select group of 'exceptional' women (whatever that word means), but about and for us all.

The second target group is anyone who is concerned to understand or develop their opinions on such women's experiences. This may particularly be people who work alongside them, or organize company systems which affect and shape their working lives and possible career development. Thirdly, and finally, this book questions how we as a society cope with change. It therefore contributes to a tide of literature which suggests that we are at a developmental crossroads, and can better shape our future if we pay more attention to the rich complexity of our present.

The main risk in writing, about which I feel uncomfortable, is that by declaring an interest in women's issues I shall immediately be labelled a 'women's libber' by many, and dismissed with all the stylized reactions this label usually provokes. That would deny me the opportunity to speak out. Please, therefore, reader, suspend your stereotypes for a while, or at least note critically those which spring to mind. For my part, I shall take some evasive action. By often presenting complexity, contradictions, uncertainties, and alternative perspectives, I shall slightly confound the process of too-easy-labelling. If you must pigeon-hole me, however, you must; and I shall learn not to mind much. I am probably more at risk, anyway, from people who do not actually read the book — stereotyping is easier the less information someone has.

Looking forward rather than backward

One choice in writing was how much attention to pay to the past. Many writers are now making valuable contributions to women's history. Amongst other things, their analyses help free us from assumptions that current standards are 'natural' or immutable, and I value this continuing work. In the main I have preferred to start from now and look to the future. I have not dipped over-deeply into why and how men and women arrived at their current relationships, distribution of roles, characteristics and relative power, but have tried to understand these as the base from which we move forward. Part of that understanding, if it is to be a useful foundation for achieving change, must be an appreciation of the functions served by particular arrangements, and of the forces and values holding them in place. So 'traditionally' applied to a skill or role in these pages means 'relatively recently' rather than 'since time began'.

There is one slant on history I do, however, find relevant to this volume, within the limits I have set. This is an appreciation of women's recent attempts to alter their position in society — and the limited impact they have achieved. Following each apparent 'success' — gaining the vote, and experiencing independence and responsibility outside the home during two world wars, for example — women have soon lost ground on the broad front of equal opportunities. Oakley (1981) explains this recurrent process as the re-assertion of male dominance in fresh guises to redress a period in which it had been brought into question. Her conclusions have significant implication for any expectations we develop of the future. The Women's Movement's previous setbacks bear testimony to the resilience of unequal relations between men and women. Appreciating that underlying forces might belie surface appearances, I was prompted to dig deeply into the book's topics, as a means of finding my way past superficial explanations. Only if we can reach fundamental influences, and address them in our actions, can we have any faith in achieving fundamental change.

Should I focus on women?

One of the more interesting dilemmas I faced in preparing to write was how woman-centred to be. Most of the available literature in the management field advocates comparative male–female studies and analyses. I became dissatisfied with this approach, because men invariably become the standard against which women's experiences are evaluated. Difficulties women experience are often discounted by saying 'its no different for men', for example. I certainly could not see the logic in allowing something which appears negative in a wider frame of values — like having a lifestyle prone to coronary heart disease — to go unquestioned because men also do it. In contrast, approaches more common amongst women, such as giving current job satisfaction priority over career development, are cast as inherently inadequate, although they could be interpreted as positive given alternative values.

I decided to focus without apology on women's experiences, in the belief that they are valuable in their own right. I particularly want to avoid merely adding women on to a world which has so far mainly been interpreted through men's eyes. I do, however, make comparisons with men as seems appropriate. In the sections which review other people's writing, being woman-centred proved especially difficult. I therefore spend some time freeing myself from seeing male as a norm, by exploring underlying assumptions about what is 'good' and 'bad'.

A BOOK PLAN, AND FINAL INTRODUCTORY REMARKS

This book, then, sets out to understand more about women in relation to management jobs. My initial mission was to speak for the women managers I interviewed — to make sense of their experiences and communicate them richly to you. As I explored further, I needed to understand more and more related issues to appreciate their experiences in context. My path has therefore been to gain illumination from several interlocking perspectives. At different times I take individual, group, organizational, structural and societal themes as my focus. In a round-about way, this further journey leads me back to a conclusion the women managers had already proposed in a simple statement — that it is very much, still, 'a man's world', and that some women can make it into management jobs, but often incur costs in doing so.

In presenting these various perspectives I take on the roles of enquirer, mediator, critic, translator and sense maker. The structure and concerns of my personal journey provide the structure and concerns for this book. A priority in offering them to you is to portray the various materials fully, to allow you access where possible to make sense for yourself — which may be a different sense from mine.

Chapters 2 and 3 were set in train by the apparently simple question 'Do women have management potential?' I found I had to explore ever more deep-seated issues about the values society attributes to being male and female, to appreciate fully the complex web of relevant factors. These chapters represent a struggle on my part to disentangle women from values about who they should be, attributed to them by men. Only towards their close do I come close to my stated intention of being able to appreciate women as themselves. Chapter 4 shifts to an organizational viewpoint, providing a complement to the previous individual orientation. It asks what places there are for women within organizations. It particularly reveals the mechanisms which keep women in low-level jobs, and which can disable then when they move into non-traditional and more senior positions.

Having set the scene with a detailed exploration of the literature, in Chapter 5 I introduce my own research study, and continue in Chapters 6 to 8 to report the experiences of 30 women managers organized around the three key themes which emerged in their accounts. Chapter 6 focuses on phases in their employment histories, and their expectations of the future. Chapter 7 covers issues of personal identity, including the management style with which they felt most comfortable. Chapter 8 looks at their overall lifestyles, and the values and practical issues these express.

The book's three final chapters are concerned with possible futures. Chapter 9 maps the standard advice to organizations and individuals on how to get more women into management jobs. Chapter 10 evaluates these strategies, and concludes that they have limited potential. Chapter 11 explores an alternative course. This is based in a broader view of the issues in context, particularly examining the relevance of societal trends in values to aspiring women managers.

Some readers may want to move immediately to the research material, and the conclusions about women managers' lives I draw from it. They should start at Chapter 5. Later, should they wish, they can work backwards to find out where my framework for interpretation came from conceptually.

Throughout the book one of my main concerns has been sensitivity to the assumptions and values underpinning particular perspectives, research findings and recommendations for action. Often I shall simply draw attention to them as significant issues. Evaluations of 'right' or 'wrong' are usually inappropriate. Many writers, for example, work from implicit beliefs that what men do or are is better *per se* than what women do or are. A few feminist authors have a reverse, but no less dogmatic, bias. Other orientations of which I became suspicious were stereotyping, about which Chapter 2 has more to say, and the use of false dichotomies which cast different qualities — such as logic and intuition — as conflicting opposites, and so effectively conceal the possibility that they might be used in effective combination.

Paying attention to values has been a provocative but often disorienting and tiring process. It has seemed like peeling back layer after layer of assumptions. Often I have been initially triumphant at identifying one bias, only to recognize a myriad of others behind it — and behind my own orientation. I do not, however, believe in objectivity. What I have sought to achieve is a critical awareness of my own personal perspective, a knowing subjectivity. A key ingredient in my biases is informing the more traditional management literature I encountered with new values which take women's experience as their base. In this sense I accept the label 'feminist' — as long as it is used to denote a community of shared concern, rather than as a standardized ideology, which it is not. Offering this perspective is this book's special contribution. Whilst my values seed these pages, they are not about right answers. My aim throughout is to provide a glimpse of options in a framework of critical, but hopefully creative, questioning — from which you and I may both choose.

Chapter 2

Do women have management potential?

This chapter and the next consider whether women are suitable material for positions in management. I explored a great deal of literature at this stage in my journey— much more than I need to share with the reader— looking for clues about women's potential as managers. I have arranged this chapter around the main reasons writers and researchers give for the relative scarcity of women managers, particularly at senior organizational levels. These reasons provide the rationale for much of the available literature. Taking each in turn— I have called them 'propositions'— I present typical research evidence to evaluate its validity. I ask how well-founded the reasoning is, and whether, if it taps a kernel of truth, the factors which prevent women's full participation in organizational life are amenable to change. From this exercise I formulate a 'response' to the initial reasoning. The six propositions I consider are:

(1) Women are different from men, so they do not make good managers.
(2) Women do not have the same motivations towards work as do men.
(3) Stereotypes of women mean companies are reluctant to employ them as managers.
(4) Women believe the stereotypes too, and behave accordingly.
(5) Other people will not work for or with women, or make life difficult for them if they have to.
(6) When women go out to work their children and husbands and homes suffer, and society suffers as a result.

These propositions do not form quite the neat, logical sequence I at first gave them credit for. In my initial explorations, as soon as I had explored one and dismissed it (as I always did) as inadequate justification for women not being, or at least becoming, managers, another would pop up to take its place. Looking at them again, later ones turn out not to be 'new' reasons at all, but old ones in disguise.

I was initially surprised to find so much research and speculation devoted to whether women are capable of doing management jobs. I had thought this question would provide a very short chapter. In fact it has spilled over into a second. The material leads me on a tour of other people's concerns rather than ones I would have generated myself. Often I did not share their initial opinions, or agree with their conclusions. I started out quite cautious of the research material and its foundations, and I became more doubting as time went on. In writing, therefore, I have two voices. One tries to understand, and adequately represent, the literature and research findings, recognizing that many people think them important. The other is continually questioning the assumptions they make, the values they express, the conclusions they draw.

This phase of my personal journey eventually led to a dead-end. I found I had been constrained by the reformist principles from which its literature emerged. My struggle to disentangle women, and myself, had been doomed to failure. Looking back I realize that all along I knew more than I thought I did. Reading only clarified my questioning, it did not move me forward. But only after several twists and turns did I explode with a full recognition of what I had continually come so close to saying. Chapter 2 ends with this new clarity, paving the way for fresh approaches to the same initial question of women's capabilities in Chapter 3.

PROPOSITION 1: WOMEN ARE DIFFERENT FROM MEN, SO THEY DO NOT MAKE GOOD MANAGERS

I was surprised at the popularity of studies asking whether women are *really* different from men. I immediately wanted to give the approach's assumptions a poke. Women, it seems, are acceptable management material if they are the same as men, unsuitable if they differ. Dissimilarity thus provides sufficient justification for excluding them from responsible organizational roles. Men become the barely disguised norm against which women's management potential is judged. Researchers do attempt some detachment by comparing various dimensions of individual difference such as leadership style or personal characteristics. But here they are on shaky ground, as there are no acceptable guidelines for what makes anyone a manager — let alone a 'good' manager. As previous experience will mainly, if not solely, be of male managers, men's characteristics again, but less explicitly, become the standard of comparison.

Given these basic assumptions, and the vital issues at stake, I could better understand the plethora of studies comparing women with men. I joined in, temporarily, wondering how similar and how different are women and men. The area of potential male–female difference which has attracted most attention in relation to women managers is that of leadership or management style. This has obvious and direct translations into performance at work.

Leadership style

Historically, and schematically, theories of leadership have developed from taking a narrow view in which the personal characteristics of the leader are all-important (so-called *trait* theories); to *style* theories which focus more on how the individual behaves and what priorities they read into the role; to more complex *contingency* or *fit* approaches which match various aspects of leader, leader behaviour, task, group and context, and evaluate their compatibility and relevance for particular purposes.

Women manager studies have favoured style theories as their conceptual base. These distinguish between two dimensions of leadership behaviour:

Structuring: A concern for the task and its accomplishment; typically associated with more directive management behaviours, rules, procedures and time deadlines.

Supporting (or consideration): A concern for people and the development of their capabilities, and for maintaining good relationships within the work group; typically associated with more participative management.

The two dimensions are seen as independent, and a typical profile using this framework would assess an individual as high or low on each — see, for example, Blake and Mouton's (1966) 'managerial grid'. The suitability of different combinations of structuring and supporting for various purposes have formed the bases of several popular leadership theories and subsequent training programmes (e.g. Reddin, 1970). Reviewers of leadership research conclude that structuring is more often associated with higher productivity, and supporting with greater job satisfaction amongst subordinates, although the associations are by no means clear-cut.

Various research approaches have been used to evaluate women managers on these dimensions. They range from laboratory studies of small groups, to surveys in which managers and their subordinates are asked to rate the manager's behaviour in these terms. I was unable to find any studies in which superiors do the rating. I suppose this is consistent with what is essentially a downward-looking model of manager behaviour, but as superiors are more likely to affect the individual's access to development and promotion opportunities, this seems a significant and conspicuous gap.

From their different perspectives, these studies come to remarkably consistent conclusions: that women are very similar to men in their leadership style. This is particularly true if we look at women already moving towards careers in management. I shall not report the detail of these results, and direct you to the, apparently more patient, authors mentioned below for their finer points. Typical conclusions from studies or review articles are:

'In most cases, there are either no differences or relatively minor differences between male and female leaders on leadership style, whether the leaders are being described by themselves or being described by their subordinates.' (Bartol, 1978; review article)

'No research has shown (sex) differences in leadership aptitudes or styles.' (Ferber and Spitze, 1979; review article)

'Women are no less qualified psychologically for positions in management than men.' (Tkach, 1980)

'Though inconsistencies abound, leader gender has generally been shown to be an important explanatory variable in laboratory studies but not in studies conducted in field settings (Osborn and Vicars, 1976) . . . Findings of sex differences in work settings disappear when the influence of age, education, and experience of leaders and subordinates is controlled (Osborn and Vicars, 1976); when type of occupation, level within the organisation, and extent of professional training are considered (Bartol, 1976; Brief and Oliver, 1976; Renwick and Tosi, 1978); and when actual rather than perceived leader behaviours are examined (Field and Caldwell, 1979).' (Riger and Calligan, 1980)

'In conclusion, there is considerable research evidence to support the fact that women mangers psychologically are not significantly different from their male counterparts and that they may possess even superior attributes and skills in some areas relating to management effectiveness. . . . Differences do exist, but mostly in ways that would serve to *increase* the probability of women functioning well as managers.' (Reif *et al.*, 1975)

Bray (1976) offers further evidence to support this strong 'no difference' trend. He draws on a review of various assessment centre programmes carried out by Bell Systems in the United States between 1958 and 1974. Assessment centres are used extensively by the company to evaluate potential new recruits, and to assess current employees for promotion. Candidates participate in a variety of individual and group simulations, designed to evoke 'managerial behaviour'. Their performance is observed and evaluated by trained staff. Results are usually fed back to candidates as well as being used as a basis for decision-making. During this fourteen year period, 132,507 participants (54 per cent of them women) were evaluated on their potential for different management target levels. Overall success rates against these criteria were 30 per cent for women and 33 per cent for men. Bray considers the difference between the two rates insignificant, given the large numbers covered. 'Character assessments' carried out by the same procedures gave parallel results. The evaluations represented good predictions of subsequent behaviour and progress.

Some leadership studies do find ways in which women differ from men, but almost always within a total profile dominated by similarity, and these differences could be interpreted as advantages. Muldrow and Bayton (1979), for example, found that women took fewer risks in decision-making, but that their overall decision accuracy matched that of the men studied. The most frequently reported difference is that women managers sometimes score higher on the supporting dimension of leadership than do male colleagues. In a review of relevant studies, Denmark (1977) concludes that women show greater concern for relationships. Some writers argue that women are suitable management material *because* of this emphasis. Interpersonal skills are portrayed in management literature as previously neglected aspects of the managerial role, to which attention and training effort must now be devoted. Women, could, therefore, be interpreted as having an advantage through their more 'natural' access to these skills of the future.

Some of the variations which are found amongst participants in management style research are clearly to do with factors other than sex differences, complicating supposed findings further. The nature of the job to be done, and relative success so far within the organization, are two of the more obvious influences.

Bachtold (1976) used Cattell's popular Sixteen Personality Factors Questionnaire to study 863 successful women representing four occupational groups. The conclusion was that differences between the groups were simply consistent with professional role expectations. Psychologists were more flexible, liberal and accepting; scientists more serious, reserved and tough minded; artists and writers more affected by feelings, spontaneous and natural, and inclined to follow their own urges; and politicians more sociable, conscientious, self controlled and group dependent.

Other authors have concluded that successful managers of whatever sex are simply different from the rest of the population. Tkach (1980) reports a study which showed relative uniformity between the profiles of 88 very senior managers (half of whom were men) in terms of work values, personality, work motivations and biographical data. The only significant sex differences reported were in lifestyle: women managers were less likely to marry, or to remarry once divorced, than were male colleagues.

These two studies' results warned me, firstly, not to evaluate male–female differences using large undifferentiated samples, as differences within groups will confuse comparisons between them; and, secondly, that comparisons should be linked to independent criteria of relevance wherever possible. They do not, however, take away my earlier suspicion that the norms of what it is to be a psychologist, a successful manager or whatever are based largely on male behaviour. These suspicions are further supported by the implications I found others drawing from the above research findings.

Despite my reservations about using male behaviour as a norm, I felt pleased and satisfied at this stage. Women do in fact pass the test by proving similar to men in terms of leadership style, and therefore suitable to be managers. I could leave my questioning to other arenas. An appropriate response to Proposition 1 runs:

Response 1: *Women are much the same as men, or very similar. Their differences are qualities you say we need more of in the future. Perhaps women will make better managers than do men.*

Further complications

But I soon discovered that even the leadership debate does not stop there. The above reasoning is too simplistic, naïve and insufficient, at least for the present, to guarantee women acceptance as managers. Having acknowledged the research findings, many writers in effect say 'Yes, but . . .'. They then go on to draw completely opposite conclusions to mine. Their qualifications pave the way for several of the other proposed reasons for women managers' scarcity explored later in the chapter. I shall briefly summarize their three most significant points.

Firstly, it is argued, women may espouse a satisfactory leadership style (and be seen by detached observers to use it in contrived laboratory tasks), but other factors in the wider context prevent them putting the style into practice. Other people's negative reactions to women in authority are particularly expected to undermine their ability to perform well.

Secondly, they say, the model of leadership on which most studies are based is inadequate. It fails to appreciate that there is more to management than these two dimensions of behaviour, and that managers operate within a complex organizational, and sometimes public, setting. Expanded views of the management role are now gaining greater acceptance and currency. Minzberg (1973), for example, identifies ten roles that managers perform. 'Leadership' in the narrow sense of relationships with subordinates is the basis for only two of these roles (which reflect a distinction between its interpersonal and informational aspects). Other (equally important) responsibilities include representing ones group within the organization; acting as an organization spokesperson; monitoring and deploying information; and performing ceremonial duties. This model incorporates an appreciation of the importance of the context to a leader's, and their team's functioning. Unless a leader is successful in organization-oriented activities — such as politics — his or her behaviour in direct relationships with subordinates may be of little consequence. Local credibility is, to some extent, dependent on organizational influence. Women's 'capabilities' in these other arenas therefore come under scrutiny. In this wider setting, stereotypes become more important as influences on behaviour—both of the woman manager, and those she deals with. Propositions 3, 4 and 5 explore various aspects of social stereotypes which impact on women's potential for management.

The third 'flaw' in my earlier conclusion that women are very similar to men hinges on the place of the supporting dimension in the style model of leadership. This is the more important dimension here, because it is the one on which women are *sometimes* found to score higher than men. Various clues suggest that this is more a handicap than a benefit to them— despite rhetoric that these are the skills of the future. Firstly, consideration is interpreted, in practice, as less intrinsically essential (to organizational functioning) than structuring because it does not have a clearly demonstrated connection to productivity. Secondly, studies suggest that consideration is judged differently according to the total profile of which it is a part. Inderlied and Powell (1979) found that their mixed-sex sample, who had varying degrees of management experience, showed such a strong preference for a male leader that this influenced their ratings of the two leadership dimensions. Other studies support their conclusion, that structuring and supporting are seen differently according to whether they are displayed by a female or male boss (Bartol and Butterfield,

1976; Petty and Lee, 1975; Rosen and Jerdee, 1973). Men are typically seen as structuring: their leadership status is accepted whether or not they also demonstrate consideration. There is no evidence, at least in studies of *women* managers, that high consideration is an adequate base for leadership. Perhaps, in practice, the two leadership dimensions are more interactive than independent. Concern for people may only become appropriate (and acceptable to subordinates) once attention to the task has established credibility and defined a context for interaction.

These tangled findings suggest that leadership characteristics and the masculine sex role correspond so closely that they are simply different labels for the same concept. Proposition 3 takes these ideas further. They also leave me wondering how appropriate it is anyway to apply the style model of leadership to women managers, as it seems so thoroughly male in its roots and explanations.

These three qualifications to Response 1 suggest that, despite theoretical leadership profiles similar to those of men, placed in the context of others' expectations and a wider web of responsibilities, women may be unable to exercise formal authority fully. Later sections of the chapter take these complicating factors as their starting points. Before moving on to consider them in detail, I shall explore a proposition which is often twinned with the one I have been considering — that of expected differences in motivations towards work.

PROPOSITION 2: WOMEN DO NOT HAVE THE SAME MOTIVATIONS TOWARDS WORK AS DO MEN

Several interrelated issues underpin research into women's motivations towards work. One pervasive strand finds it strange that women should want to work at all, and sets out to understand their reasons for doing so. (This perspective assumes as strongly that men do want to work, as that women do not.) Once their interest in working has been established, a second strand of enquiry tries to make sense of, and evaluate, the various reasons women give. The third debate follows from the second, and is particularly relevant to management jobs. It asks whether women want jobs or careers.

It has long been assumed that women differ from men on all three aspects: that they are not committed to paid employment, work for less serious reasons than men, and prefer to do a job rather than develop a career. All three assumptions have been used as arguments that women are unsuitable for responsible jobs, and as sufficient explanations of the low-level organizational places they occupy. I shall consider the issues in sequence.

Do women want to work?

It is becoming compellingly obvious that women do want paid employment. They can no longer be considered a secondary labour pool which takes up jobs when they are plentiful or in times of national emergency, and returns to the home when no longer needed. We are currently experiencing a test of women's determination in this area which shows its strength. As the economic recession bites deeper, the traditional opinion that women are not entitled to take jobs from other people is being voiced increasingly strongly.

Just over a third of the employed labour force in Britain is female. Oakley (1981) corrects impressions that the proportion of working women is increasing dramatically. They were 31 per cent of the employed labour force in 1850, 1911 and 1951 alike. The recent increase is relatively small. The major change has, in fact, been in the number of *married* women who work. In 1911, one in ten married women had a job; in 1951, one in five; in 1976, one in two. Two out of three employed women are married (Oakley, 1981). This reflects many factors, including the trend to smaller families and the subsequent decline in time spent child-bearing and rearing. Its significance here is the major shift in attitudes it represents. Marriage and work are no longer the mutually exclusive options they once were. Given the opportunity, many women want to participate meaningfully in both worlds.

In an early study in the area, Hunt (1968) captured the force of women's interest in work. Her data came from a General Population Survey of the United Kingdom in 1965. Seven thousand women were interviewed to provide a detailed report on the 'who, why and how' of working women. It is interesting now to see just how much about their participation in employment appeared at the time to need explaining. Hunt notes with interest, for example, that having the first child rather than marriage had become women's main reason for leaving paid employment. The sample's scores on the study's 'work-mindedness scale' (being employed or definitely intending to return to work when possible) show a consistent interest in having a job. 'Work-mindedness' was about 80 per cent amongst the 16–24 year old age group, and hovered steadily around 70 per cent for those between 25 and 49 years old. Amongst older age categories it dropped off steeply, reaching only (but hardly surprisingly) 37 per cent for those between 60 and 64.

The report goes on to illustrate why the sample were not more involved in work in practice. Significant constraints, which limited their choice of job and their day-to-day performance, were: strong feelings of responsibility for child, home and husband care; their husband's attitudes (although 17 per cent of the women who worked said they did so despite disapproval); local availability of employment; and lack of appropriate training. Given this mix of competing demands, it was hardly surprising that 65 per cent of those who worked said it had been important to them to get a job within easy travelling distance of home. Hunt concluded that working women were 'wholly or mainly responsible for running their homes', and that 'help' from husbands was limited to washing up. Current studies of who does the housework reach surprisingly similar conclusions (Oakley, 1981; Gowler and Legge, 1982). Even when household activities are a little more evenly distributed, responsibility for housekeeping still tends overwhelmingly to rest with the female partner.

Women's interest in employment may, then, appear less substantial than it is because other factors either prevent them from working altogether, or from taking the kind of job they would like. I now turn to the motivations they satisfy through working.

Evaluating women's work motivations

In terms of specific attitudes, a recurrent issue is whether women are really serious, and motivated (in the ways that men are) about their jobs. Hunt's (1968) sample certainly had low expectations of gaining satisfaction from the job itself, or of achieving career progression. Only 20 per cent gave the use of their skills or qualifications as motivations for working. Obviously differences within the sample (in terms of education and past training, particularly) become important here. The main reasons given for working, by members of the sample who had jobs, were:

(a) money, either as essential family income or to afford luxuries (mentioned by 80 per cent)
(b) pleasant working companions (65 per cent)
(c) good working conditions (59 per cent)
(d) desire for company (40 per cent)
(e) to escape boredom (40 per cent)

(Most did not have a single reason). These results illustrate the profile of motivations traditionally ascribed to women workers. Their priorities were expected to be money — for family rather than personal use — and friendship. As general statements, such stereotypes are now out-of-date. More recent studies tend to find few significant differences between women and men in work motivations, particularly when the groups studied are similar in terms of other key features such as occupation, organizational level and education. Whilst the samples in many of these studies are small, the consistency of their findings offers a significant counterweight to traditional stereotypes, which they most assuredly contradict.

A typical comparative study of motivation at work is that of Brief and Oliver (1976). One hundred and five retail sales managers (of whom 53 were women) indicated their expectations of meeting sales targets, and the importance to them of 25 job outcomes such as fringe benefits, working relationships, prestige, responsibility, job security and personal growth. (The items were drawn from Vroom's model of motivation (Vroom, 1973)). Holding constant the potentially confounding variables of occupation and organization level, the authors found no significant pattern of male–female differences.

The literature exploring why women work is drenched in interesting assumptions about what constitute legitimate motivations. The evaluations represented amount to a powerful double-bind for the woman worker. Women are traditionally excluded from management jobs because they are judged less serious, less highly motivated than male employees. In particular they are supposed to demonstrate low organizational commitment because they do not assign their jobs precedence over all other life areas, may leave to have children, and demonstrate less company loyalty than do male colleagues. Paradoxically, working tends to be considered socially acceptable only if it can be portrayed as essential to one of their other life roles. Wanting to earn money to contribute to their family's survival or well-being, for example, is relatively understandable. There is even a hint of scandal, however, in reports that married women's wages are not required for necessities, but provide the family with 'luxuries'. For the first time in 1980, the Family Expenditure Survey analysed the effects of a wife's wages on family spending patterns (a noteworthy exercise in itself). Its results were worthy of publicity because the greatest impact appeared to be increased consumption of holidays, alcohol and tobacco. Women seeking employment to enjoy the job itself is depicted as even less legitimate.

In public, this double-bind is revealed in the explanations people are prepared to give for their behaviour. It is safer for a working woman to say her family needs the money. Employers or professional bodies proposing retraining schemes refer to the women concerned as potentially wasted, expensively trained, resources. The use of such stylized justifications deters women from expressing more personal motivations, and helps perpetuate inadequate stereotypes.

For the woman manager the double-bind represents a delicate balance of factors. If she shows too little interest in work, she will tend not to be taken seriously. Although she will be complying with female stereotypes, she may also be seen as a threat to organizational stability because of her apparent independence and lack of commitment. 'Commitment', however false it is in practice, is a traditional symbol of company loyalty. Women create uncertainty if they cannot be relied on to give the job priority over other life interests. If, on the other hand, the woman becomes too involved in work, she risks different forms of censure. She becomes 'real' competition to male colleagues, and is accused of neglecting home responsibilities, and her own 'femininity'.

Jobs or careers?

Attitudes become particularly polarized when we return to the old adage that men have careers whilst women 'merely' have jobs. The double-bind identified above is obviously of further relevance here. Many women do now have an interest in continuous, or nearly continuous, employment. Seventy per cent of a sample of 124 women aged between 24 and 55 in lower- and middle-management jobs plan to work until retirement age, for example (Norgaard, 1980). There seems, however, more variety in what women mean by this than I would expect for a male sample.

Two women-only studies suggest that, even if they are interested in continuous employment, women do not necessarily subscribe to traditional notions of 'career'. In this area, for the first time in this journey, I consistently found more differences than similarities between women and men. Mirides and

Cote (1980) refer to a recent survey which asked 355 women what they wanted from employment. The total sample covered a mixture of occupations. The motivations expressed by women in managerial and administrative jobs did not differ significantly from those of other groups. The first preference of 60 per cent was for meaningful work. Fifty-five per cent saw a high income as a priority. Only 15 per cent stated an interest in promotion. In their survey, Norgaard (1980) also found that money was important. But their results suggest caution in interpreting its meaning. Sixty per cent of the sample gave financial incentive as their major reason for working. Their qualifying comments reveal, however, that earning power was a source of satisfaction in itself, and not a purely instrumental motivation. To this group at least, money stood for other non-financial motivations, such as independence and identity.

Three recurrent themes appear in women's accounts, and distinguish their concerns from traditional notions of 'career'. Challenge and satisfaction in a particular job are more important than recurrent promotion for its own sake. Work is important, but does not take priority over all other life areas. Women tend not to plan far ahead (Hennig and Jardim, 1978; Ruddick and Daniels, 1977). Levinson (1978) identified an implicit idea of a career timetable in his study of an all-male sample of executives, novelists, biologists and manual workers. Women appear to have a different sense of time. They talk in terms of seeing what develops, and making the most of what happens.

In later chapters which recount the work experiences of 30 women managers, these themes will appear again. It is interesting to speculate whether their differences are because of women's general unfamiliarity with certain areas of the work environment, or whether they are expressing values significantly different from men's values. They certainly challenge the powerful norm that continuous, committed employment is both 'good' and essential to successful management performance. Discontinuous employment may have its benefits. Levinson's data suggest that long-term career plans can become too fixed and unresponsive to changed circumstances. Several of his sample who eventually achieved their 'dreams' in their fifties or sixties suffered a reactive depression, realized that their long-sought goal was not what they now needed, and gave up their work-long career paths.

In these sections on their attitudes towards work, women have again proved mainly similar to men. I have uncovered no reasons why they should be incapable of being managers. Practical difficulties of balancing several life roles may stand in their way of translating interests in work into practice. The unequal legitimacy of women's different motivations towards work leads me into more complex issues — about stereotypes initially, which are explored in the next three major sections, and later about social values. Whilst some women do seem to hold notions of career different from those of men, these do not detract from their potential as managers. These issues will receive further attention during later sections of the chapter. On the current topic of attitudes to work, I can now respond.

Response 2: *Many women have motivations towards work similar to men's. Practical difficulties explain why some do not translate these into action.*

The material on leadership styles and attitudes to work did little to explain the real differences in the distribution of women and men in organizations. If individual characteristics are an insufficient explanation, I must look elsewhere. One suggestion is that it is not what women are like, but how they are seen by others, that determines what opportunities are open to them, and how effective they will be in particular circumstances. There seems little point in asking if women are the same as men, if their similar behaviour is judged differently. The next three propositions explore various impacts of social stereotypes on women's chances of achieving, and performing well in, management jobs.

PROPOSITION 3: STEREOTYPES OF WOMEN MEAN THAT COMPANIES ARE RELUCTANT TO EMPLOY THEM AS MANAGERS

If employers' stereotypes of women do not match their requirements of a manager, the argument goes, they will not consider employing them. The most popular approach to research in this area has been to elicit stereotypes and assess the 'fit' between those of 'women' and 'managers'. The conclusion is repeatedly reached that management is stereotyped as a male occupation. Schein (1976) sums this up in the provocative title *Think Manager — Think Male*.

Since their first publication in 1973, Schein's research methods and findings have achieved the status of classics. Using a 92 item checklist, she asked 300 male middle managers in various United States insurance companies, and later 167 similar female middle managers, to describe their perceptions of 'women in general', 'men in general' or 'successful middle managers'. Both samples saw a large, and statistically significant, resemblance between 'men' and 'managers' ($r' = 0.60$ for the male and 0.54 for the female samples). The female sample reported a weaker resemblance between 'women' and 'managers' ($r' = 0.30$), and the male sample near zero correspondence between the two. Thus, for men at least, there was no 'natural' overlap between the notions 'women' and 'manager'. Massengill and Di Marco (1979) provide more recent confirmation of these findings. Table 2.1 shows the evaluations of Schein's original samples in more detail. Both samples see some similarities between female and manager in terms of 'employee-centred' characteristics, such as helpfulness, and in terms of competence and intelligence. As this suggests, most women currently in middle and senior management have entered through expertise (staff) related rather than production related functions. Women's image of being employee-centred is, however, less advantageous than it might seem, if we remember our earlier discussion of leadership style.

Schein concluded that women's route into management is fraught with difficulties. Women will be discouraged from thinking of management as a potential career because they do not identify very strongly with the role. Should

Table 2.1 How female and male managers described successful manager characteristics (adapted from Schein, 1976; reproduced by permission of the *Atlanta Economic Review*)

Requisite management characteristics which *Men* are expected to display

By female sample only:	*By both samples:*	*By male sample only:*
(Non) desire for friendship	Leadership ability	Emotionally stable
	Competitive	Steady
	Self-confident	Analytical ability
	Objective	Logical
	Aggressive	Consistent
	Forceful	Well informed
	Ambitious	
	Desiring responsibility	

Requisite management characteristics which *Women* are expected to display

By female sample only:	*By both samples:*	*By male sample only:*
Modest	Intuitive	Sophisticated
Creative	Helpful	
Cheerful	Humanitarian values	
	Aware of feelings of others	

Requisite management characteristics which *either Women or Men* are expected to display

By female sample only:	*By both samples:*	*By male sample only:*
Emotionally stable	Competent	Tactful
Steady	Intelligent	Curious
Analytical ability	Persistent	
Logical	(Not) devious	
Well informed	(Not) bitter	

they apply, they will find entry difficult because (male) employers do not see them as suitable material. Those who are accepted will have difficulty achieving a viable self-image and style, as the management stereotype is so consistently male. Ironically, if they adapt to prevailing norms as a personal survival strategy, women managers are likely to be no more tolerant of other females at work than their male colleagues are. On a slightly more hopeful note, Schein reports that the resemblance of manager to male was most strongly expressed by women with limited management experience. This suggests women go through an initial identification phase, in which they copy men to prove acceptable. By doing so they earn the freedom to develop their more female characteristics and skills — unless (as it may well have done) socialization has inhibited their ability to do so.

One of my reservations about Schein's studies is their use of such broad categories as 'men in general' and 'women in general', as their initial reference points. They do prove the point beautifully that managers are more likely to be male, but then I already knew that. Perhaps the enthusiastic acceptance of the findings by the academic and popular press, despite the small and particular samples on which they are based (another methodological weakness), bears testimony to the power with which they do so. They tap 'the kernel of truth' behind the stereotypes, as Allport (1954) would say. What they fail to do is to explore our ability to discriminate within these gross generalizations, or more importantly, the extent to which we do so in practice. Other researchers have found, for example, that people rate themselves as less sex-typed than the general stereotypes they describe. We are also learning more about the complexity, diversity and real nature of managers' work, making 'successful middle managers' a uselessly broad and idealized category too. I would prefer to see researchers dig more deeply, and be prepared to communicate complex findings, rather than promote simplifications.

Schein's results, and those of similar studies, do have their realm of validity. They capture stereotypes by demonstrating the close correspondence of male and manager in both women's and men's eyes, and the relative incompatibility of female and manager. They are not sufficient justification for excluding women from management jobs because, as previous sections have demonstrated, the stereotypes do not adequately represent women's characteristics and capabilities. Rationally, then, I can argue that employers should accept the research proof, change their negative opinions of women's suitability, and recruit them to management jobs.

Response 3: *But we know stereotypes are ill-founded. They are not an adequate excuse for excluding women from management.*

Stereotypes are not, however, so easily dispensed with. A further proposition approaches the same material from a different direction, and reminds me of their impact on the behaviour of women themselves.

PROPOSITION 4: WOMEN BELIEVE THE STEREOTYPES AND BEHAVE ACCORDINGLY

This proposition appears to be well-supported by research and other evidence, and is one of the more forcefully argued explanations (in its various forms) of women's scarcity in management jobs. It is captured in the occasional use of the phrase 'women are their own worst enemies'. My earlier case seems to be undermined if women themselves believe, and behave in accordance with, non-managerial stereotypes. To understand potential women managers in these terms, I need to take a major detour into the nature, power and implications of sex role stereotypes. In doing so I come nearer to disentangling women from the simplified images society holds of them. This process involves two complementary strategies. One is a continual awareness of the use of male behaviour as a norm against which women are judged. The other is a willingness to ask what is wrong with what women do — even if it is different.

The nature of sex role stereotypes

Sex role stereotypes are essentially social creations. They are the meaning assigned to being biologically female or male. They have their roots in anatomy, but are much more besides. Much of the teaching we receive during childhood socialization instructs us in sex role requirements (Sharpe, 1978). Direct injunctions, such as 'little boys don't cry', and 'that's not a very nice thing for a little girl to do', are the more obvious tip of a largely insidious iceberg of shaping and moulding. This process has a profound impact on core aspects of individual personality. As gender is a more central foundation of identity than most (if not all) other characteristics, sex roles have a commensurately significant influence on who we are, how we behave, how others see us, and how others behave towards us. Our sex role 'permeates all aspects of life, and takes precedence over other, more situation-specific work or social roles if they are incompatible' (Bayes and Newton, 1978).

'Masculine' and feminine' are the usual labels for sex roles, paralleling 'male' and 'female' which are used to distinguish biological sex. A typical picture of masculinity involves such designations as: able, achievement-oriented, active, adventurous, aggressive, assertive, authoritative, bold, breadwinning, competent, competitive, consistent, dominant, engaged in the public world, hard, in control, independent, intellectual, money-oriented, organizing, powerful, rational, realistic, status-conscious, strong and unemotional. Commonly accepted notions of femininity provide a thoroughly different characterization. A typical list would include: cautious, changeable, concerned about physical appearance and age, co-operative, delicate, dependent, domestic, emotional, emphasizing relationships, fertile, fragile, generous, gentle, inconsistent, intuitive, kind, loving, nurturing, passive, patient, personal, receptive, responsive, sentimental, showing concern for others, soft, subordinant, supportive, timid and weak.

Society's heros and heroines reflect these profiles' contrasts. Our heros are rugged footballers, successful businessmen, bold adventurers. Our heroines are supportive wives, warm mothers, caring nurses, sexually attractive women, innocent maidens. Men's ideal roles seem more easily expressed as the roles themselves, women's tend to emerge in terms of their relationship, or potential relationship, to other people. A distinction which is often proposed as a reliable sex difference is that men are self-sufficient for their experience of personal identity, but that women rely on relationships for their sense of self.

In making the above comments, I have been swift to relate to the stereotypes, find them interesting and make sense of them. Unintentionally I have illustrated my own next point: that however aware I am that the stereotype is a poor representation of the world as I know it, I all-too-soon use it, invest it with meaning, and thus make it real in its consequences. It has been impossible to write without generalizations. Preparing the following sections has acted as a salutary warning to handle stereotypes with caution, and take what opportunities I can to check them against alternative, more specific, perspectives.

Sex roles as potential traps

Sex roles are normative. They provide helpful guidelines about appropriate behaviour. They also act as constraints to individual freedom. Some role aspects are defined more prescriptively than others. I do, for example, feel free to wear trousers to many social functions, but would attract far more attention for non-conformity if I smoked a pipe. As individuals, we will also vary in the necessity we attach to different sex role norms.

Much of the literature about sex roles emphasizes their potential to stifle expressions of individuality and to constrain growth and development by their demands. A Canadian report on the different uses to which men and women put Valium (a widely used, tranquilizer), provides an interesting insight into sex role differences, and the pressures they represent (Cooperstock and Lennard, 1979). People's continuing use of the drug could only be understood 'in terms of "permitting" them to maintain themselves in a role or roles which they found difficult or intolerable without the drug'. Men's problems were the pressures of being in paid employment. Women needed help to tolerate their domestic roles, and the conflicts between competing demands. They particularly experienced stress at failing to meet traditional female role expectations, and feeling that they should not express overt dissatisfaction, and yet could not escape the situation. One mother said: 'I take it (Valium) to protect the family from my irritability'. Even children, it seems, show more awareness of disadvantages than advantages of their sex roles. Vincent (1978) asked school girls and boys to write about themselves and each other. Stereotypes had changed very little during ten years of Women's Movement activity in the United Kingdom. Here are just some of the children's complaints.

The boys It's not fair because we have to do all the dirty jobs and the ladies have the clean jobs.' . . . 'What I don't like about being male is that when you cry your parents say: "Come on, you're a man and men don't cry". But when it's your sister they're all heart. Also, if you put on something that smells nice, girls will say you are a pansy. They think the ideal man should be big and good looking.' . . . 'I don't like being a boy because when I'm ill my mother says: "come on get up, there's nothing wrong with you", but if it's my sister she's sympathetic. If I hit my sister I get framed as a bully, but when she hits me I'm supposed to be tough and walk away as if it didn't hurt.'

The girls 'I hate the way women are always put down, the mother-in-law, the wife, women drivers, these are all joked about. We have to do all the manual tasks and our brothers just do nothing. When you advertise for cleaners it's a woman who answers because her husband is more than likely drinking in the public house.' . . . 'You can always hear about women being attacked on the streets but a woman attacking men is unheard of. The future doesn't exactly look bleak but I certainly will not be a housewife. I want to work in the world.'

Interestingly, mothers were often portrayed as the chief teachers and enforcers of these discriminating sex roles. If anything, the girls did see more potential flexibility in their roles, expressing an opening up of possibilities which seems to reflect the current questioning of women's roles and capabilities. For example: 'Boys are supposed to play rougher games but some girls are rougher than some boys. Some women, for instance, are gunwomen (terrorists).'

An extreme response to feeling trapped in one's sex role may be to change one's biological sex altogether. Eichler (1980) suggests that sex-change operations are one of the more spectacular repercussions of narrowly defined sex roles. She draws this proposition from studying accounts written by transexuals and their doctors.

Eichler interprets seeking a sex-change operation as a sign that the applicants cannot achieve important life goals within their given bodies. Ironically, these goals appear to be related more to psychological and social, than to anatomical, needs. Eichler quotes a follow-up study by Bentler (1976), of 42 male-to-female, post-operative transsexuals. The primary reason the sample gave for changing their sex was 'to make my body more like my mind, as a woman'. This was chosen above such alternatives (in a multiple-choice questionnaire) as 'wearing pretty clothes', 'being less aggressive', and 'having sex with a male'. From this, and other data, Bentler concludes that surgery was seen as the only way to allow expression of psychological characteristics, and was relatively less important as a means to fulfilling physical sexual needs. This material suggests that transsexuals have such overdetermined, and narrowly defined, ideas about what men and women do, they have to change their anatomy to earn the behavioural freedom they wish. The doctors they consult appear to be working from similarly narrow views of sex-appropriate behaviour.

If doctors can be accused of perpetuating and over-rigidifying sex role stereotypes, so too can social scientists. One of the ironies in this area of study is that the over-definition and polarization of stereotypes has been significantly fostered by researchers, many of whom would doubtless declare rather different intentions. The construction and use of masculinity–femininity scales in behavioural research is a prime example of their role in defining, fixing and perpetuating sex role stereotypes (Eichler, 1980). By using specifically requested sex role generalizations as initial data; by selecting for inclusion in the 'test' only those characteristics on which significant sexual differences can be elicited (ignoring many similarities); and by using resulting scales as idealized norms in further studies, researchers allow initial assumptions of male–female difference to become self-fulfilling prophesies through the process of research.

In other male–female comparative studies, bias is often shown in the analysis and reporting of findings. Characteristics which men and women share are deliberately muted, they are not the 'news' reporters are looking for. Instead differences (even those at very poor levels of statistical significance) are highlighted and explored at length. Foster and Kolinko (1979), for example, compare women who chose a 'non-traditional' university course (Master of Business Administration), with those doing a traditional women's subject

(Education) on various psychological tests. They discuss the four scales on which the two samples differ, virtually ignore seventeen further factors on which they show no significant differences, and sweep on to conclude that 'Today's woman preparing for a career in management appears to be a special breed when contrasted with her more traditional sisters'. I am not convinced.

Sex role stereotypes are, then, as much potential traps for researchers, professional helpers and writers, as they are for people generally. This is particularly important to me here, for its effects on research — which has an image of providing objective data about the world, but really does no such thing. Each study is bounded by the assumptions its originators bring to their work. If they assume success in management for women is reaching the top as often as men do, they will count the number of women at senior organizational levels and report this as a percentage in relation to men. They have already narrowed their field of enquiry dramatically — particularly by excluding alternative goals (and combinations of goals) people may pursue through work, and by using men as the standard by which to judge what women do. Their assumptions will owe a great deal to cultural norms.

As a consequence, I find many of the studies reported in this chapter more informative because of the interests and assumptions they express, and the *nature* of the conclusions they draw, than for their supposed findings. These typically have a narrow range of applicability because of the constraints of the researchers' original assumptions. In this sense the studies are useful as evidence of themselves rather than as the evidence about an actual world their originators intended.

Femininity — what does it mean

As women are my main concern, I shall now probe more deeply into their sex role — labelled femininity. In doing so, its two levels of entrapment become more apparent. Not only are the norms of what it is to be feminine constraining, the notion itself is illusory, bearing little resemblance to women's actual experiences of themselves and their role. To explore these issues with any vitality and penetration, I found myself drawn to a more radical literature than I had previously tapped. It answered more of my questions, and was soon to provide an escape route to a more meaningful phase of exploration.

In a more detailed examination of the notion of femininity, a significant difference between it and masculinity shouts for attention. The concepts are not equivalent because they are not assigned equal social worth. Masculinity and femininity make up a societal structure in which femininity is consistently one-down. Its characteristics — such as emotionality, domesticity and nurturing — are perpetually devalued compared with masculine traits of rationality, public action and independence. As a result, the female sex role itself is revealed as a devalued social role. The motives that identify its characteristics, and set it below the masculine role, express male rather than female interests.

One of the main signs of femininity's lack of female grounding as a concept is its failure to capture the totality of women's experience by acknowledging its negative, painful aspects. 'The point is, of course, that male made-up femininity has nothing to do with women. Drag queens, whether divine or human, belong to the Men's Association' (Daly, 1979). The media's glowing image of mother-hood is a particularly clear example, which consistently ignores the pain, toil, boredom and frustration which are also part of that role. If, despite this idealization, women accept stereotypes of femininity they have to struggle to meet their expectations. 'In a male culture, the idea of the feminine is expressed, defined and perceived by the male as a *condition* of being female, while for the female it is seen as an *addition* to ones femaleness and a status to be achieved.' (Cornillon, 1972). Without sustained attention my feminity might, in effect, lapse. Characteristics or experiences which fail to match the stereotype must also be repressed or denied. In this sense Daly (1979) criticises both feminine and masculine as 'sets of characteristics which are essentially distorted and destructive to the self and to her process and environment.' Novarra (1980) says she has never understood what femininity means. Is it, she asks ingenuously, 'a genteel way of saying sex appeal?'

Another strand of evidence that current sex role definitions have more to offer men than women, comes from appreciating the covenience of the tradi-tional division of labour between the sexes as a way of getting society's tedious, less glamorous work done. Here, yet again, I encountered a curious paradox which places women in a double-bind about their worth. An implicit societal theme (too diffuse to ascribe to particular sources) implies that women's traditional roles— particularly those of taking care of home, husband, children and the sick — support the social fabric. It finds expression, however, in confusingly negative ways. Women are not openly acknowledged and por-trayed as holding the community together, and rewarded accordingly with social value. Rather, their contribution is honoured in the breech— women in employment are accused of neglecting their duties. Being expected to fulfil roles for the common good, which are paradoxically ascribed a low social status, creates a difficult double-bind, which is mirrored in women's conflicts about their personal, as well as their social, worth.

On examination, femininity as a social role allows an individual less self-determination and fewer personal rights than masculinity does. It carries within itself injunctions which prevent the individual both from appreciating that it is a trap, and from making sense of experiences which are outside its narrow boundaries. I shall show in later chapters, however, that realization of femin-inity's potential to constrain and distort is growing. Increasing numbers of women are resisting the sex role's directives, and are looking for new notions of identity.

My exploration of femininity and masculinity left me thoroughly dissatisfied with the unsoundness of their conceptual foundations, and their ambivalence in practice. I considered abandoning them altogether — but relented when I realized that doing so without a viable alternative base would risk substituting new labels meaning the same things. I found I still needed to use them, or at least allow them a place in my thinking. They are both consciously 'real' to other people — so women managers may evaluate themselves, and be evaluated by others, against notions of femininity — and have become part of the social norms and fabric of society, influencing our unconscious perceptions of fit, appropriateness, possibility and the like. The notions of feminine and masculine are therefore tolerated but not welcomed in the pages which follow.

How women fulfil notions of femininity

Some writers have become quite excited by the realization that women apply (negative) stereotypes to themselves and each other, and often behave accordingly. This is obviously important, but is not surprising given the power of socialization, and of the social processes through which sex role norms are enforced. Neither is it sufficient proof that the stereotype is valid or that women are unfit to be managers.

Several themes emerge from studies which show women apparently conforming to notions of femininity, illuminating the usefulness and limitations of research in the area. Some studies illustrate the potential costs of the 'deviance' of disconfirming sex roles. Many reveal an independence in women's thinking and values which is worthy of further attention. Of the three areas of research I have chosen to illustrate these issues, the first two depict women as part-victims of stereotypes — at least in public. All three do, however, show them offering alternative perspectives which challenge basic assumptions of desirable management behaviour.

Leadership and visibility

A now classic study (quoted in Denmark, 1977) shows how important it may be to people to maintain a stereotype, at least at the level of appearances. Individuals were pre-tested with questionnaires, and rated for 'dominance'. Pairs consisting of one high- and one low-dominance person, were then asked to choose a 'leader', as preparation for taking part in an experiment. High-dominance women tended to end up as 'followers' in 80 per cent of such pairs — but they had made the decision to do so. As it is largely unacceptable for women to take the lead, particularly if that involves giving orders to a man, these women were showing an appropriate awareness of, and responsiveness to, social norms. From this I could conclude that women are unable to perform well as managers. If I look more closely, however, I find that women high in dominance did exert influence, and achieve an outcome which suited their social purposes of not appearing deviant. Their use of power was rather different from that typical of male leaders — but nonetheless effective.

Non-verbal behaviour and relative status

My second example of the constraints women may experience through sex roles reaches to more fundamental aspects of behaviour. Video recordings have been used to explore body language in relations between male and female managers (Denmark, 1977). Males consistently behaved dominantly, and females submissively in their non-verbal behaviour. From an extensive review of the available research, Henley (1977) concludes that this is the typical pattern in male–female exchanges and reflects differences in social power. High status is associated with smiling less, initiating more touching, acting in a relaxed manner, and not adjusting ones behaviour to other people's dominance initiatives. In contrast to this, the women studied tended to increase affiliation and minimize distance. They also adjusted their behaviour to others. This behaviour was partly dependent on the setting. Differences in social status were more significant in certain contexts. If women communicate non-verbal messages of submission, these contradict their verbal statements of leadership, and undermine their ability to exercise formal authority. Researchers consistently find that if verbal and non-verbal communications contradict each other, more attention is paid to the latter. Non-verbal behaviour may then be more significant in communicating status than is sex *per se*. The values on which these judgements of status are based are obviously open to question. Maintaining distance may not be *the* (one and only) way to be a manager. The power of the current norms and expectations they indicate cannot, however, be ignored.

Sex differences in explanations of success

One of the many illustrations of the application of double standards to female and male managers is the different explanations given for successful performance. Similar behaviour is consistently found to be judged differently, by both observers and the managers concerned, according to whether it is displayed by a woman or a man.

Whilst observers of laboratory studies (doing anagrams and the like) attribute a man's success to innate ability, they are more likely to see a woman's as the result either of chance, the simplicity of the task, or, possibly, of effort (Deaux, 1979). Ability as an explanation implies personal merit, and is likely to attract rewards of increased pay, opportunities and promotion. The latter cluster, in contrast, focuses on situational or transitory factors, and tends not to count towards individual credit (Heilmar and Guzzo, 1978). Deaux found that performers parallel these evaluations, in their own. Women rate their performances less highly, tend to use fewer internal, personal, reasons to explain success, and use more less permanent explanations than do men. If (with due caution given their limited research base) these effects are transferred to the workplace, women can be expected to make less of their successes, and draw less learning from them to guide future behaviour, than do male colleagues.

Field studies support the above findings (Riger and Galligan, 1980), and suggest that stereotypes affect evaluations most when success is ambiguous. If independent criteria of performance are available, the effects can disappear altogether.

The accumulating literature in this area testifies to how long-term and pervasive can be the distrust of their own abilities that women learn through socialization. Some people appear to retain their poor self-image despite outstanding occupational achievements. Clance and Imes (1978) met the 150 highly successful academic and professional women they studied whilst the latter were engaged in some form of personal growth training or therapy. All considered themselves 'impostors', and lacked the internal sense of success their degrees, prizes, professional recognition and job appointments might be expected to confer. Many were convinced that they were fooling other people into thinking them intelligent, and would eventually be discovered. The women's strategies for coping with their fears tended to maintain the impostor phenomenon. Some, for example, hid their own opinions or used personal charm as a basis for work relatioships. They then worried that had others known their true thoughts or personality, they would be exposed as the frauds they feared they were.

I shall end this section, as I do many others, with a cautionary note. These studies suggest that women are less able than men to perform well as managers. But I am suspicious of the values on which such judgements rest. Yet again it is insidiously appealing to use what men do as a standard for 'good' managerial behaviour. The very format of research does so. As someone concerned about women's own capabilities, I prefer to explore alternative interpretations. There may be value in not boasting about what one does, or even not claiming 'success' on an individual basis. Similarly, approaching new situations with few preconceived ideas or action strategies can have advantages.

Recent research findings also teach me to be continually open to revising what I think I know, in the light of evidence that values are changing. Garland (1977) provides an interesting illustration of sex stereotypes losing some of their relevance. This experimental study using American college graduates as subjects is more valuable for its comparative findings than for the originality, or worth, of its research methodology. Participants read descriptive profiles of different people, and responded by attributing them with characteristics on various rating scales. Similar studies in 1975 had generated wholly sex-linked findings. Then, any males were evaluated positively, any females negatively. By 1977, sex took a back seat. Instead, workers who had been successful were viewed more positively, and expected to experience more positive consequences in the future, than were unsuccessful workers — regardless of sex or occupation. Success was attributed more to ability and effort, and less to the job itself, than was failure.

Deviance from sex role norms: fear of success

Because of their core significance for identity, deviating from sex role stereotypes can have significant social and personal consequences. Non-conformity attracts disapproval and sanctions from others — such as ridicule or social isolation — which the individual may find disturbing. We also judge ourselves against our own internalized, but not always internally consistent, norms of acceptable behaviour. Breaking, or failing to live up to, norms of femininity can result in self-punishment and guilt. Much of stereotypes' powerful controlling effect on behaviour comes from individuals expecting and trying to avoid the consequences of deviation. A further complication for women is that they do not always have to do something to transgress norms of femininity. As these have constantly to be maintained and reaffirmed, inactivity, too, is potentially deviant. In many ways, then, I see the above studies which show women conforming to sex role stereotypes as evidence of women being kept in place by social norms. Even when they used independent values of their own, the women were partially restricted because these were misinterpreted as part of a, largely negative, female stereotype.

I shall use one final illustration which brings together deviance and management jobs before formulating a 'response' to Proposition 4.

A chapter on the impact of personal factors on women's potential for management would be incomplete without some consideration of 'fear of success', a factor proposed, and warmly accepted, in the late 1960s as an adequate explanation of women's lack of career progress. In essence, it is argued, women are afraid of transgressing against sex role norms by performing well in traditionally male roles.

Horner (1972), the notion's originator, suggested that women avoid success as it is in opposition to, and therefore jeopardizes, their femininity. The study on which she based these conclusions asked students to write a story that began: 'At the end of first term finals, Anne finds herself at the top of her medical class'. (Horner starts, then, with a very male view of 'success'.) Females wrote stories about 'Anne', males about 'John'. Analysing the stories for themes, she found that those of the women were consistently concerned with the negative consequences of success. They expected to be socially rejected and lose their friends; expressed fear and other negative feelings; gave hostile responses, or denied the 'facts' of the original cue. Horner (1972) concluded that:

'Among women, the anticipation of success, especially against a male competitor, poses a threat to the sense of femininity and self-esteem and serves as a potential basis for becoming socially rejected . . . in other words, the anticipation of success is anxiety producing and as such inhibits otherwise positive achievement motivation and behaviour. In order to feel or appear more feminine, women, especially those high in fear of success, disguise their abilities and withdraw from the mainstream of thought, activism, and achievement in our society . . .'

'Fear of success' seems originally to have been interpreted as an individual fault. More recent commentators have argued that, taking social norms into account, the respondents' concerns are very realistic. Tresemer (1974) therefore prefers the label 'fear of sex role inappropriateness'. This new slant is supported by men's apparent fear of success in what for them would be unusual occupational roles. Male executives asked to finish a story which departs significantly from expectations for them — 'at the end of first year finals, Paul finds himself at the head of his nursing school class' — express considerable anxiety and usually reject the basic premise that Paul really intends to be a nurse (Bardwick, 1979).

Consistent with this new slant, further research reveals the importance of contextual influences on women's attitudes to success. Women in management jobs who have found their place within the norms of that culture write realistic stories about the repercussions of good performance, based on their own experiences of being successful. They focus on the logistics of time pressures or of managing home and work responsibilities, for example. Bardwick claims that the women in middle management positions she works with (in her roles as researcher, trainer and organization consultant) cannot be induced to give fear-of-success responses — however hard she tries! Unlike Horner's original sample, they are not expecting circumstances of which they know little. As yet, too, their success is not noteably deviant, being in ocupational fields, and at organizational levels, at which women are well represented.

In this substantial detour into sex roles, I have portrayed their power to shape and constrain individuals' behaviour. Women may behave in apparent accordance with stereotypes for a variety of reasons. I have found that they particularly like to sustain stereotypes in terms of appearances (although their behaviour may belie them in other ways), and that they do express alternative, independent values of their own. Whatever their intention, however, other people's perceptions seem largely stereotype-bound. This has been brought home most clearly to me in the normative interpretations researchers build into their studies and take from their findings. They move between using female and male stereotypes as standards. Women are censored for conforming to images of femininity — they give different explanations to men of successful performance — and for failing to match characteristics of masculinity — not wanting to be top of the medical class.

This detour has also taught me that I do not have to cling to the narrow views of the world in which these notions of sex inequality are enshrined. Stereotypes are socially created and sustained, and are not, therefore, indestructible. Women appear particularly sensitive to other people's expectations of them — if these changed, their sex role would be less constraining. They would also be able to develop the independent perspectives and values they express, enriching the work environment as well as their own lives.

Response 4: *Stereotypes trap women and men. But we create them so we can also change them. If attitudes to women changed, their views of themselves, and their behaviour, would change too.*

I seem only to be learning one thing repeatedly through the course of this chapter: that unless women behave very similarly to men they will not be considered suitable as managers. But from an exploration of stereotypes, I have discovered something more besides: that if they stray too far from stereotypes of femininity they will be sanctioned for deviance. Fulfilling the stereotype is how they become recognized as women. Current equal-opportunities legislation represents a self-defeating paradox to the extent that it ignores these powerful stereotyping effects. Legislation officially gives women the opportunity to be equal to men on men's standards. It does not allow questioning of current norms, or foster growth towards a jointly meaningful, mixed-sex public world. As a result a part of the total social system — women's place — is meant to change without significant accommodations and transformations of that system as a whole. This is a logical, practical and existential impossibility. In particular, women seeking to be 'equal' to men in the world of work and public affairs will continually find themselves in conflict with men's traditional expectations of femininity, and their interest in enforcing these.

At this stage in my researches, I was faced with a personal dilemma. My initial enthusiasm had waned. Each time I explored one objection to women being managers, and found it insufficient, another would pop up to take its place. Two things particularly sapped my energy. I seemed to be encountering very similar roots behind each proposition. I wanted to move forward, rather than cover the same ground repeatedly. Secondly, my responses had begun to sound increasingly hollow because each new wave of research material showed that, however sensible, the 'truths' they tap have a limited impact in practice. For one last time I responded, and moved on to the next proposition, which had at least sparked my interest. But I did so uncertain of my path.

PROPOSITION 5: ATTITUDES HAVE NOT CHANGED SUBSTANTIALLY —OTHER PEOPLE BELIEVE THE STEREOTYPES, AND DO NOT WANT TO WORK WITH OR FOR WOMEN; IF THEY HAVE TO, THEY MAKE THE WOMEN'S LIVES DIFFICULT

One of the reservoirs of openly expressed reluctance to employing women as managers is the claim that 'other people' do not want to work for, or deal with, women at work. Harvard Business Review surveyed 2000 subscribers (half of them men) in 1965 (Bowman et al., 1965) on their attitudes towards women executives. More than two-thirds of the men and almost one-fifth of the women said they would feel uncomfortable working for a female boss.

From a more recent study of 1400 male and female academic employees and university staff in a large mid-Western American university, Ferber et al. (1979) derive a similar picture of preferences. Respondents were asked how they would react to women either as bosses, or in the six professional occupations of accountant, dentist, lawyer, physician, realtor (estate agent) and vet. Table 2.2 summarizes their replies.

One conclusion the authors drew from their findings was that women were not their own 'worst enemies' — men were, by a small margin. Their survey does, however, suggest that attitudes towards female bosses or professionals

Table 2.2   Attitudes towards the sex of boss or professional worker

	Males	Females
Indifferent	26%	30%
Prefer male in at least one occupation	64%	51%
Prefer female in at least one occupation	17%	30%

are changing. Greater acceptance was associated with exposure to women in these roles, with higher education and (for men), with being married to working women. They conclude that affirmative action legislation will have a significant impact simply by increasing the number of women in management positions. The authors also explored how experience influences stereotypes. Unless the increased number of women are seen as competent, it seems, they will create more negative attitudes towards women (amongst men), and will not therefore have a generally beneficial effect. Experience of competent or incompetent men was not generalized to influence stereotypes of male managers. Presumably the initial stereotype was relatively well-established in the first place. Exposure to competent women was generalized, by both sexes, to improve their image of women's capabilities. Men also generalized any experience of incompetent women to develop negative stereotypes.

A valuable illustration of some of the above themes (and others from earlier in the chapter) in action comes from a case study of a newly appointed department head, and how she fared in the job (Bayes and Newton, 1978). The study is about an American Community Health Centre team, and its newly appointed, and first female, head — Dr. A. One of the authors acted as a long-term consultant to the team, and the article provides a rich account of its development and concerns. Stereotyping issues and aspects of management style are portrayed in the context of organization structure and staff group dynamics (topics which receive attention in Chapter 4). The account demonstrates the importance of how a leader is seen by members of the team, how they behave towards her, and her ability to liaise with the wider organization, as determinants of her overall effectiveness. The case demonstrates how inappropriate it is to look for a single cause as an explanation of such a complex web of influences.

Dr. A experienced problems managing both the team's external and internal boundaries. Her ties with the total organization were difficult; she was largely excluded from informal relations with the all-male group at her level. As a consequence her team became more isolated. Some members became particularly anxious about this, and tried to initiate liaison with Head Office independently. Because of their poor relationship, the group was less protected from intrusion by superiors. Outsiders were allowed into departmental meetings, for example, and this was interpreted by subordinates as another sign that Dr. A was not doing her managerial job properly.

Within the team, subordinates seemed uncomfortable responding to Dr. A's authority. They tried to attract a more distant (male) superior into taking an interest in the team. When this failed, they tried to cast Dr. A in the role of assistant to other (male) figures — including the consultant at one stage. Challenges to Dr. A's authority were masked and unacknowledged. They came in the form of covert defiance or denial of subordination, for example, and were therefore difficult to identify and address.

In this context Dr. A's reactions often seemed weak, and inappropriate. The subtlety of the challenges to her authority robbed her of direct feedback on subordinates' concerns, and of a pretext on which to express her authority openly, which she generally seemed hesitant to do. She relied throughout on a supportive style which failed to improve her situation. Her attempts to enrol a series of male team members as direct assistants were also rejected. As a result, Dr. A was at increasing risk either of isolation, or of drowning in detail by trying to do too much unassisted, both of which possibilities would have made her situation even more difficult.

A further strand to the profile is the strong dependency needs having a woman leader seemed to stimulate in the group. In trying to handle the 'greedy' demands made of her, Dr. A was faced with equally dangerous alternatives of acquiescing to the role and trying to satisfy them, or of failing to do so and thus risking further sanctions from her staff. (An observational study by Mayes (1979) also found that women in authority elicited hostility or dependence in men with whom they work.)

Bayes and Newton identify Dr. A's inadequate use of power, team members' dependence and lack of previous experience of a woman as leader, and the inadequacies of her reactions to difficult situations, as the key factors which interferred with Dr. A's ability to do the job of team leader well. They conclude that

'a woman in authority should be prepared to counteract strong social forces in herself and in others which act to preclude competent leadership behaviour. An understanding of such potential difficulties helps a work group and its female leader to mobilise resources in more effective ways.'

The case study of Dr. A and her team leads me back again to the conclusion that how a woman is seen and treated is far more important in determining her management potential than are the personal characteristics she brings to the role. It goes further, however, and allows some detailed insights into what this means in practice. How strongly relations between subordinates and leaders, and between colleagues (members of the senior management team at Head Office), are based in norms of, and a preference for, maleness is a repeated theme. Dr. A was perceived from the outset as a leader manqué. One of the case's most striking aspects is the contribution the team made to disempowering Dr. A. Their behaviour continually confounded any potential she had to operate successfully. Even disappointment about her performance as a manager could not be expressed productively. Dr. A's own impact on the situation was partly substantive, and partly symbolic. Faced with difficult situations, she became trapped in stereotyped roles. Bayes and Newton suggest she could have acted differently, and drawn on more behavioural options. How possible being independent of stereotypes is, and what values such independence can draw on for its foundations, are key issues in this book. The case material shows that Dr. A was not only 'not male', but was symbolically female. The expectations and needs this raised for subordinates represented another level of entrapment, which Dr. A fostered by persisting with her supportive management style. Simply maintaining her 'femininity' is not, therefore, an adequate solution for the female manager's dilemmas.

Any response to Proposition 5 would be very similar to those of its predecessors: suggesting that we adapt stereotypes to appreciate women's demonstrated capabilities, and that if this were to happen women would be able to deploy their skills more effectively. I have also identified the next, sixth, proposed explanation or justification of women's exclusion from management jobs. It is showing up in studies which report organizational policy-makers as favourably disposed overall towards recruiting women as managers. They recognize, I am told, that women and men are equally qualified, but still express some reservations. According to surveys by Rosen et al. (1975) and Vandermerwe (1979) these lingering doubts return me to sex roles yet again. Potential employers feel that:

(1) women allow their emotions to influence their management behaviour;
(2) the possibility of pregnancy reduces women's value as employees;
(3) women make better mothers if at home;
(4) women will be unable to balance work and family demands;
(5) women will not get the support of their husbands for working.

PROPOSITION 6: WHEN WOMEN GO OUT TO WORK, THEIR CHILDREN, HUSBANDS AND HOMES SUFFER, AND SOCIETY SUFFERS AS A RESULT

I could engage in rational debate about these topics, quote studies which fail to find any negative effects on children whose mothers go out to work, criticize the myths about maternal deprivation which took such a hold in the 1950s and 60s, and look at the poor mental health of women who mind their homes reluctantly. *But I shall not, because I have had enough.*

I understand now that whatever I say, someone will tell me that it is not quite proof enough. Whatever women do, it will never be 'right'. They are striving to meet criteria which are repeatedly changed just as they seem within reach. Women are also faced with a series of double-binds, which essentially encourage them to copy men's characteristics and behaviour to become acceptable managers, but punish them for departures from stereotypes of female behaviour. My journey so far has been like travelling across shifting sands. Although I feel I have worked hard — reviewing the relevant material, making sense of lists of sometimes conflicting conclusions, and moving on each time along the trail as apparently new issues emerge — I am no further forward with my initial quest. As one apparently rational argument is discredited and I think I have reached firm ground, this gives way as the original argument is replaced by one equally plausible. In time, this reasoning too can be contested, but the ground shifts again, and another proposition takes its place. Underneath all these apparently rational requirements there are more fundamental principles show-ing their shapes. Each proposition has the same effect of excluding women from management, but draws on a slightly different proof.

Figure 2.1 documents the unsteady steps I have taken so far. Legislation may actually contribute to this bizarre process, by chasing issues underground as it outlaws open expressions of prejudice against women. My journey suggests that the fundamental issues do not go away; they simply emerge in more socially acceptable, disguised forms.

PROPOSITIONS	RESPONSES
1. Women are different from men, so they do not make good managers.	1. Women are much the same as men, or very similar. Their differences are skills you say we need more of in the future. Perhaps women will make better managers than do men.
2. Women don't have the same motivations towards work as do men.	2. Many do. Practical difficulties stand in the way of others.
3. Stereotypes of women mean companies are reluctant to employ them as managers.	3. But we know stereotypes are ill-founded. They are not an adequate excuse for excluding women from management.
4. Women believe the stereotypes too, and behave accordingly.	4. Stereotypes trap women and men. But we create and sustain them. They aren't indestructible.
5. Other people won't work for or with women, or make life difficult for them if they have to.	5. If we gave women more freedom, they would demonstrate their abilities. Stereotypes would change and relationships get easier.
6. But when women go out to work their children and husbands and homes suffer, and society suffers as a result.	6. No they don't! But what is all this?
	EXIT!

Figure 2.1   Standard propositions explaining why more women are not managers — and some responses

Halting, I find that so far I have been working on other people's ground, addressing issues they originated. The burden of proof and justitification has been on my shoulders, and 'proof' has seemed called for by the very nature of the frameworks used. I have been on the defensive. I realize, too, that I have been sustaining those frameworks, by taking the propositions seriously. My odd shots and questionings have been influenced by their concepts. I may have dented their assumptions, but my triumphs have been mainly adversarial — destructive of potential enemies, but not constructive for myself. The greatest irony of my part in this process is that I had not set out wanting to prove women the same as men in the first place. I was lulled by finding them similar on certain accepted measures of leadership potential into thinking this would assure their initial acceptance, and that notions of productive dissimilarity could find their place later. But none of the above propositions let me escape the repeated injunction that men are the standard for whatever women do — and that men are the judges too. These insidious norms consistently assign women a low status. Having at last realized that my oblique attempts have done nothing to weaken this injunction's power, my next step becomes obvious. I must look directly at its workings and foundations; at why it is less 'good' to be a woman than it is to be a man, and at what parts women and men play in the processes of assigning these social values. Chapter 3 provides me with a fresh start in these new directions.

Chapter 3

Social values, power and women

I: MY CHANGING ORIENTATION: FROM REFORM TO RADICAL FEMINISM

This chapter emerges from my dissatisfaction with the material which has gone before rather than from its content. Its sense is therefore, dialectic rather than sequential. I felt I had still learnt so little about women's potential to be managers, that I had to shift to a new level of enquiry, one which would encompass all the topics above at once, and more. In Chapter 2, whatever women did they could never be 'right'. I therefore decided to look at how we decide what is 'right', as this process seemed to be largely independent of logic. Two topics provide the core for this new phase of discovery: how society assigns worth or value to some people, characteristics and activities and not to others; and social power which determines whose values gain general acceptance.

I also had to find a new base of literature and ideas to turn to. By the close of Chapter 2 I was thoroughly dissatisfied with the material I had been tapping. It contained too many values and assumptions which displeased me. I had not, either, particularly enjoyed this stage of my journey. It was like travelling along a darkened tunnel, with only occasional flashes of excitement or relevance. Looking back, I see this as my 'reform feminist' phase. I was viewing women from the perspective of established norms and systems, finding out how they do or do not 'fit'. As the female viewer, I deplored women's exclusion from the potentially exciting world of public action. But the frameworks I was borrowing captured so little of my own experience that my reaction was mainly of opposition. I often felt passive, muted frustration at writers' implicit assumptions, but could not say clearly what I wanted instead. Eventually my dissatisfaction became so great that I had to break out of this lethargic tunnel. In doing so I found new positive directions just waiting to be taken. There was another strand to women's thinking which I had so far hardly touched on. [Partly, I now realize, because it did not appear on the computer search printout which had been keeping me so busy].

This chapter taps these new sources, which I now identify as those of radical feminism. I found a whole new realm of ideas which affirmed my earlier dissatisfaction, and helped me explain it in broad, far-reaching frameworks. These perspectives took women, rather than men, as their focus, and so resonated with the core of my own experience, which the reform feminist ideas had either ignored or misinterpreted. This new material welcomed me, seeming to know who I was. Much of the writing had the quality of expressing with clarity what I knew, inarticulately, as 'truth'. As my reading and thinking touched my personal sphere in this powerful way, I experienced a new sense of myself — as a woman, and with new freedom. I began the delicate, engrossing process of developing my own feminism.

The literature of radical feminism into which I plunged at this time covers a broad span of approaches. It incorporates some widely shared beliefs, but no one ideology is identifiable, or sought. There are several reasons for this, some relating to the diverse historical and political backgrounds from which those who call themselves feminists come. Three factors particularly dictate that feminism is an individualistic rather than a collective matter. Firstly, personal experience is honoured, and becomes the standard against which to gauge 'truth'. Individuals must look to themselves, rather than to external authority to develop and evaluate understanding. Following on from this emphasis is an appreciation that there are different perspectives on the world rather than a verifiable objectivity. This second factor sets the scene for respecting diversity in attitudes and beliefs. It is also significant that feminism is an emerging and evolving consciousness. It has not yet reached, and I hope never will reach, a plateau of understanding.

Within feminism I found a respect for female experience. What women do, say and think is taken as the foundation on which to base concepts and models of the world. This in itself is unusual, and hence 'radical'. (Although from now I shall drop the label 'radical' unless wanting to make a distinction with the reform approach. I do so for simplicity, and because, for me as a woman, these ideas have become a wholly 'natural' form of feminism). Feminist writers point out that men are more usually the centre of public attention, whilst women's experiences are dismissed as trivial. Many have therefore devoted their energies to disentangling the value judgements these sex differences reflect. Western society is essentially a patriarchy in which men hold authority, and women are either oppressed or ignored. These inequalities are reflected in our social institutions, in the distribution of work, in the pattern of interpersonal relationships and the language through which we express and transmit our culture. Women are consistently in a one-down position relative to men. As men benefit from women's social inferiority, they are seen as both passively and actively perpetuating it through various mechanisms, which range from the social level of excluding women from responsible jobs, to the interpersonal level of making demands on women in the service role of housewife.

Much of the feminist literature reaches out towards new futures in which women are both socially valued, and freed from the constraints of recent sex roles. Authors vary in the demands they consequently make of men. Most see the first major step as being women's own valuing of themselves and each other. Some feel that women must also separate themselves off from men's female-destructive influence, if only temporarily, to achieve their own development unhindered.

This phase of my journey held the mingled delight and shock of riding a fairground helter-skelter. I seldom knew where I was going next, or how steep the slope would be. It had significant landmarks, such as particularly relevant, illuminating, disturbing or personally meaningful concepts or writings, and plenty of those precious times when ideas collide and new sense is made. These form the skeleton of this chapter. The ride was exhilarating and engrossing, but had a depressing, alienating side too. Some of the ideas made sense to me, but so upset my previous views of the world that I hardly knew how to behave. As I developed an understanding of the deep structure of language, for example, and the ways it consistently devalues female in relation to male, all the conceptual distinctions I was so used to making were brought into question. I weighed every action and word, and was often dissatisfied with myself. I had to scrutinize and then re-work much of what I had previously taken for granted.

Underneath all this turmoil, a strong sense of increasing knowing and personal worth was growing, as I did battle with so many of my own assumptions, and survived. I found I could embrace challenges and uncertainty with a faith in my own competence and opinions. My sense of humour returned. I developed, too, a clearer appreciation of the range of beliefs and experiences encompassed within feminism.

Surveying the field now I can choose my place within the radical feminist frameworks which dominated this phase of my journey. I am still, however, a little cautious. Whilst feminism explicitly offers freedom, some of its literature is as normative, evaluative and condemning of alternatives as the belief systems it denounces. Encountering such attitudes I become fiercely independent. I will not risk being swept along again by a perspective which denies me my sense of the world. Within feminism there is, however, sufficient latitude and welcoming of new enquiry for me to escape such a fate. As a broad area of thought and writing it is the nearest to 'home' I have found on my journey. From passive, frustrated opposition, I moved to rewarding, active exploration. This chapter presents some of the highlights of that exploration — selectively and briefly, because there were so many.

As I move forward, I feel that the tone of my writing has changed. It is no longer doggedly attacking, but accepts and seeks to build on the ideas presented. In this section I have described experientially the change in belief and consciousness which gave me a new starting point. I shall not argue its case logically through the material below, because that would be inappropriate. I shall simply allow my revised viewpoint to speak out through my explorations. The remainder of this chapter covers the following topics:

II: Social valuing processes reflected in language

My first concern is how social values are arrived at and maintained. In this section I explore language as a carrier of values, to learn about the supposed relative merits and faults of being female. I encounter a pattern of social inequality, in which men have supremacy.

III: Towards explanations of social power

I next look for an explanation of sex structuring in society. I want to understand how male dominance operates. An appreciation in these terms is essential before we can attempt to change social norms towards valuing women. Through this analysis, I arrive at my own model of female and male qualities in a framework of potential equality and relationship. This is the book's key theoretical offering, and seeds later interpretations of managers' lives and proposals for the new directions women managers can take.

IV: Women's characteristics and culture

At last I feel I have achieved a new perspective on women, relatively free of social stereotypes. In this section I therefore gather together a profile of women's characteristics, the resources on which they can draw to develop their skills as managers.

V: Androgyny, and adding shape to communion's combinations with agency

The chapter's last major section briefly discusses androgyny, a notion to which some writers are now looking to resolve conflicts about sex roles and relative values. In some ways their ideas parallel my own, in others we differ radically. I explore the merits of androgyny as a concept, and in the process learn more about my own framework of communion and agency.

VI: Concluding remarks

Finally, I summarize the implications this phase in my journey has for the book's development.

II: SOCIAL VALUING PROCESSES REFLECTED IN LANGUAGE

In Chapter 2, I often called into question how we evaluate 'good' and 'bad', 'suitable for management jobs' or 'unsuitable', and expressed dissatisfaction with the criteria on which such judgments are made. Turning my analysis to the social processes by which worth or value are assigned takes it to the heart of such matters. As valuing is a largely unconscious process, I shall approach it through illustration. I shall show that, as norms stand, male is consistently associated with superior and positive and female with inferior and negative. Of many available options, I have chosen language as my example of valuing in practice. Language is a fundamental element of culture: reflecting, perpetuating and shaping values and consciousness. In '1984' Orwell (1954) reminded us of its power:

> 'The purpose of Newspeak was not only to provide a medium of expression for the world-view and mental habits proper to the devotees of Ingsoc, but to make all other modes of thought impossible. . . . a heretical thought . . . should be literally unthinkable. Its vocabulary was so constructed as to give exact and often very subtle expression to every meaning that a Party member could properly wish to express, while excluding all other meanings and also the possibility of arriving at them by indirect methods. This was done partly by the invention of new words, but chiefly by eliminating undesirable words and by stripping such words as remained of unorthodox meanings.'

Language is a major cultural system which by its very shape and nature reveals the culture's key principles and underlying pattern. That pattern in western society is one of patriarchy — of male authority, involving domination of the main sources of social power. Norms of male dominance and superiority are enshrined in language just as unambiguously as they are in social institutions and public life. Exploring one system in detail illuminates the nature and workings of the others.

In this section I shall consider several ways in which sexual inequality in social power is revealed in both the structure of language and the nature of conversation. My main concerns are the repercussions for women's places in society and for individual women's experiences. A fuller review of language *per se* can be found in Spender (1980). This book was one of the reading highlights of my journey for its scholarship, ability to speak out from female experience, and sense of balance.

Male as the norm

A fundamental principle of language which reflects and expresses male superiority is the taking of male as norm. This is more readily apparent in some guises than in others. The standard, widespread usage of 'he' and 'man' is the most obvious expression of this principle. In addition to indicating known males, 'he' and 'man' are used as generic terms for human beings, to refer to mixed sex groups and when the sex of referents is unclear of unknown. He/man may also be considered grammatically correct forms regardless of the sex of the person involved. The last rule even substitues them for she/women in the latter's only specific area of relevance — that of known female designation.

The usage of the he/man convention reveals the supremacy of maleness in the deep structure of language. This has several repercussions for women. One of the most important is that they are rendered invisible. Man becomes the accepted symbol of person — the image brought to mind is biologically male. Public expression and writing thus becomes a chronicle of man's experience; their viewpoints predominate and their needs are expressed. Women are either absent altogether from these references or their role is made unclear.

The general use of he/man is often defended by claiming that these are generic terms which include women too. This claim is refuted by current research evidence that typically both the reader and writer have male 'men' in mind. Recent studies show that whatever the writer's avowed intention, readers do not include women in their interpretation of the word 'man'. Phrases such as 'thinking men' and 'economic man' or statements such as 'an Englishman's home is his castle' tend to call up images of male people only, not female people or females and males together (Miller and Swift, 1981). Often women appear initially to be included in the supposed generic use of 'man', but as more details emerge the implicit maleness of the reference shows through. The following examples, taken from recent published material, are typical of this confused and confusing use of 'man', and the enigmatic role in which woman is therefore placed:

'As for man, he is no different from the rest. His back aches. He ruptures easily, his women have difficulties in childbirth . . .'
(quoted in Miller and Swift, 1977)

'How does Man see himself? As a salesman? A doctor? A dentist? . . . As far as sexuality goes, the Kinsey Reports on the activities of the American male surely affect his self-image in this regard . . .'
(quoted in Spender, 1980)

In both cases early usage could well include women, but its development reveals a solely male reference. Similar jolts of re-interpretation are required by much of the business studies literature. Managers are referred to as 'he', and the female reader's attempts to include herself in this designation are continually undermined by lapses which show that this 'he' is indeed male, rather than generic or simply grammatically convenient. When, as often happens, 'managers' turn out to have 'wives', for example, the illusion is quickly shattered.

Whilst working in a University library to prepare this book, I came across a telling illustration that women are a subcategory of the human race needing special designation. I also found out where they (who, I wonder?) had put us. As many libraries and bookshops are now doing, this one has recently created a new subsection of 'women's studies' to handle a growing tide of publications. This alone could be interpreted as proof that previous studies have only related male experience. Searching round for an appropriate Dewey classification number someone had found '390: Folklore, Custom' and had added to it '396: and Women'.

Our use of male as the norm has a profound impact in women's lives. Women are rendered marginal and unusual, and as a consequence inferior. 'By promoting the use of the symbol *man* at the expense of *woman* it is clear that the visibility and primacy of males is supported. We learn to see the male as the worthier, more comprehensive and superior sex and we divide and organise the world along these lines' (Spender, 1980).

Male positive, female negative

Our use of language to make distinctions of value provides a further strand in the translation of patriarchy into linguistic practice. Fundamental inequalities of social power are encoded in the basic categorizations and judgements of relative worth which constitute language. Miller and Swift (1977) use the term 'semantic polarization' for the pervasive process through which words associated with female and male are cast as opposites, and female forms consistently devalued in relation to their male counterparts. This devaluing is often associated with derogatory references of an explicitly (biological) sexual nature. The opposites we distinguish in relation to sex are not, it seems, of equivalent status, but are essentially asymmetrical. Patriarchy determines their relative worth. As a result female terms consistently emerge as negative, male terms as positive.

Titles provide one field of examples of this semantic derogation of women. Whilst *spinster* and *bachelor* both designate unmarried adults, the latter term has none of the negative connotations of the former. *Sir*, *Master* and *King* are all still used as titles of respect, whilst *Madam*, *Mistress* and *Queen* have developed debased meanings. Lakoff (1975) points out the considerable descrepancy in meanings between 'an old master' and 'an old mistress'. Our judgements about characteristics and attributes follow a similar pattern. Qualities identified as male — strength, rationality, perseverance, for example — accord with cultural norms of what is 'good'; those associated with female — yielding, emotionality, changeability — with 'bad'. Spender identifies semantic polarization as a corner stone of patriarchal language and culture: 'It is this superiority/inferiority dichotomy which is a principle encoded in our language. It is the prefigured pattern of our language which serves as the source of new names for men and women' (Spender, 1980).

It would appear that even computers are programmed with sufficient assumptions creatively to devalue women. To track down more material on women managers I requested a computer literature search including the stimulus words 'women' and 'retailing'. Back came the printout offering, amongst others, a research report from an American University entitled 'Sex for Sale, Improve Your Health: Advertisement of Prostitution Services'. I paused to reflect. It seemed that the computer was part of a cultural boomerang system — but whatever gets thrown out, the same few things come back.

Other examples show that the interaction of sex with evaluative designations is an active process of evolving meaning. Thus the new association of male or female with a designation can move it up or down the semantic status hierarchy — respectively. According to Miller and Swift, for example, 'seamster' was originally an all female occupation. When it was taken over and became dominated by men the new title 'seamstress' (with a French derived ending) was created to denote women workers. Men were thus left with the original, more prestigious, job title. A more recent case also shows how women and high status jobs are incompatible, and that circumstances shift to restore underlying inequalities if they become disturbed. As the general importance and therefore status of personnel — originally a female *function* par excellence — within organizations has risen, increasing numbers of men have moved into what is now clearly a *profession*. The numbers of women, particularly in top jobs, has correspondingly declined (Novarra, 1980). There is currently some speculation amongst those who recruit new university graduates into industry that the increased proportion of women taking business studies courses in one form or another will actually devalue the resulting qualifications.

Kanter (1977) uses the term 'status levelling' to refer to this process of adjusting one aspect of status to match another. Sex consistently represents the 'anchor' status to which other designations gravitate, with female low and male high. Whilst male nurses are misperceived as doctors — their status is adjusted upwards to match their maleness — female executives are mistaken for assistants or secretaries — femaleness adjusts their status downwards. For Spender (1980) the meaning of the consistent typing of male positive, female negative is clear: 'All words — regardless of their origin — which are associated with females acquire negative connotations, because this is a fundamental semantic 'rule' in a society which constructs male supremacy'. These principles are therefore both enshrined in, and expressive of, patriarchy itself. From the power base of patriarchy, men claim the right to construe public meanings. In exercising this 'right' they have, and do, impose their values and standards, based firmly in their experiences, on society as a whole. Language is 'man-made' in Spender's terms. Against these standards women are judged as different and therefore inferior.

Perhaps, in view of the conclusions this section is reaching, the title 'devaluing processes' would have been more appropriate for the woman-focus I claimed to be taking. But I was slow to draw that connection, and will therefore leave the current title as evidence of my myopia.

The absence of female meanings from language

As the dominant social group, men shape the language. This, however, is only part of their formative impact on the construction of knowledge more generally. Knowledge is not objective, independent of people's attempts to know, but is structured by the frameworks we bring to it: 'What we observe is not nature itself but nature exposed to our method of questioning' (Heisenberg, 1963). What western society accepts as science is a world exposed almost solely to men's questioning. Feminist analysts challenge society's classification of subject areas, identification of topics, methods of research and subsequent theoretical models because women have played no part in their construction. 'Every discipline must have a paradigm ... these paradigms which are rigorously followed (and taught) were not handed down from some benign authority: men made them up' (Spender, 1980). Women's absence from this process has contributed to their invisibility and to the ommission of their meanings — either in terms of symbols or structures of thought — from language and science. We do not have to resort to biological determinism (although I believe that biology plays a significant part) to expect that women's participation would have made a difference. Women and men are positioned differently in patriarchal order and have different social histories. On these bases alone, they generate sexually differentiated meanings and explanations. From their combined differences: 'not only do men and women view a common world from different perspectives, they view different worlds as well' (Bernard, 1973).

From their social position, women have not had the same opportunities (as men) to influence the language, to introduce new meanings where they will be taken up or to define the objects or events of the world. Women's meanings are not, therefore, encoded in the language. In their place, we typically find men's attempts to speak for and about women. I identified some of the resulting concepts in Chapter 2 as idealized stereotypes of femininity which owe far more to male conceptions of who women should be and how they should relate to men than to women's own experiences of their roles.

In speaking and writing, women are therefore using what is to them a foreign language, because it does not include direct reference points for their experience. Women must engage in a translation process to convert their meanings into male terms and forms of speech in order to express them in socially understood and accepted terms. Rich (1979) talks of 'having to tell our truths in an alien language' and Olsen (1978) of 'telling it slant' to describe this extra step which women may have to take in communicating. Women may find that there are no available words to express their meanings. Spender uses 'starting by being in the wrong' as an example of a concept which is readily understood by many women she talks to but has no appropriate label. The term 'sexual harassment' is an example of a label newly created to fill such a gap. The phrase has recently gained currency as women explain and debate the implications for them of encountering unwelcomed sexual advances. Men and women divide up the world of their experience differently (for example, Bernard, 1981). These differences too require acts of translation on women's parts if they are to use male concepts appropriately. In general men appear to make more distinctions and demarcations to map their experience, whilst women respond to wholes and patterns in their entirety. Women will also find that the shape of conversation is dictated by male conventions. Women have to suppress emotional tone and personal reference (important aspects of *their* world), if they want to be allowed to join in.

The translation process of 'telling it slant' helps make sense of several of the criticisms of style levelled at women who enter traditional male job areas. 'Not to be able to come to one's own truth or not to use it in one's writing, even when telling the truth having to "tell it slant", robs one of drive, of conviction, limits potential stature' (Olsen, 1977). These extra steps in communication contribute to women's hesitancy in public, which is often attributed instead to lack of confidence. Women will also probably find that the messages they want to communicate become distorted, both in conceptual accuracy and personal expression. They are therefore vulnerable to being misunderstood by their listeners. Whilst the frequent accusation that female managers are 'hard' reflects several contributory factors, women's needs to distort and mute their own truths in order to express themselves in acceptable managerial words and tones must certainly play a significant part.

Women also have to engage in translation processes in reading and listening. As I noted earlier they will often be unable to tell whether they are, or could be, included in communications which use he/man terminology. They will have to look for other clues, or ask supplementary questions to resolve the ambiguity. Even if women are included too, imagining themselves doing or being whatever is referred to necessitates a conceptual substitution which is not required of men. The extra activity demanded of women is only one of its costs. Women's identification with the original communication is also undermined. The female business student reading about 'the manager, he . . .' will not have the immediate impression of herself in that role that her male counterpart experiences. Her initial image will exclude her. If she does then map herself into the text in a personally meaningful way, this sense of identity will be temporary and conditional, and will have to be remade each time she re-encounters he/man terminology.

The scarcity of female meanings in language limits communications between contemporary women and across generations. Women tend not to talk to each other on a public scale or to exchange truths based in their own experience when they do so. They are thus prevented from developing a joint consciousness of their social position. Women's failure to pass on a substantial female legacy contributes to their fragmentation. Surveying reading lists and bibliographies of twentieth century literature as one measure of women's cumulative heritage, for example, Olsen (1977) found only one woman writer mentioned for every twelve men.

The realities of female experience have to be discovered anew by each successive generation in the culture. Coping with these realities is made the more painful because they conflict so markedly with the male notions of womanhood enshrined in cultural stereotypes.

The sections above have built towards a multi-stranded appreciation of the ways patriarchal order achieves women's silence and invisibility. Men also play an active role as boundary-keepers to limit women's participation in the construction and use of social meanings. The scarcity of published material by women is an example at the social level. The media and publishing companies have, until recently, been run mainly by men. As a result, the writing women have done despite the obstacles (Olsen, 1978) has generally been ignored, if not deliberately suppressed, and little has therefore passed into the culture. Spender provides a more day-to-day illustration. She chronicles ways in which women are kept out of conversations by being interrupted and ignored unless they introduce topics men want to talk about. Contrary to popular myth, in mixed company, research evidence shows that men consistently talk more than women. Men rarely listen silently to a group of women talking, whereas the reverse is a common occurrence. These differences, too, relate to power. Independently of sex, high-status individuals tend to dominate social interactions. As the superior social group, men determine and control the shape of conversation. As a result, women spend far more time listening than talking.

So far in this chapter I have illustrated men's domination of the construction of language, and women's subsequent invisibility and silence. I shall now explore three particular examples of these processes in action. All three demonstrate how men use their control of language to value themselves and devalue women.

Labelling cognitive styles

One of men's major uses of linguistic power is to assign cultural worth through the process of naming. An influential series of experiments into 'cognitive style' conducted by Witkin et al. (1962) illustrates how some individual characteristics become identified as better than others (Archer, 1978; Spender, 1980).

'Cognitive style' encompasses how we perceive and think about the world, core aspects of human functioning which writers — most noteably Jung — have used as a framework for charting and understanding individual differences. Witkin's explorations were laboratory studies which assessed whether participants saw a stimulus as separate from its surrounding contextual field, or saw the stimulus and context as a whole. In one set of experiments, for example — the tilting-room/tilting-chair test — subjects sat in a tilted chair which they were asked to adjust to the vertical, despite misleading environmental cues designed to distort their sense of upright.

An implicit assumption of the researchers' concepts and language is that separating stimulus from context is a 'better' cognitive style. People who relied more on their own sense of upright to adjust the chair were labelled 'field independent'. Those who paid more attention to environmental cues, and therefore made less pronounced adjustments in posture, were termed 'field dependent'. Sex differences were found in the style adopted: women more often emerged as 'field dependent'. Given their initial orientation this only confirmed an inferiority which had been prefigured in the (male) researchers' initial assumptions. Spender suggests the new terms 'context awareness' and 'context blindness' to reverse Witkin's labelling bias. I can leave the reader to decide how she applies them.

Work: non-work

The issue of what is work can also be dealt with in terms of labelling. The nature of work and women's place in the employed labour market are two topics which have received considerable attention recently. These debates conceal a battle about what constitutes 'real work'.

Many writers argue that western industrial society's distinction between 'women's work' and 'men's work' serves social rather than biological purposes. Cross-cultural comparisons show that it does not correspond to any inherent differences between the sexes, not even the capacity for heavy physical labour (Oakley, 1981). Designations of work, then, provide further evidence of the deep structure of valuing, and the low status assigned to women.

Novarra (1980) identifies seven tasks which traditionally constitute women's work in western society. They are:

(1) assuring continuance of the human race by bearing children;
(2) providing and preparing food;
(3) clothing people;
(4) tending sick and frail members of society;
(5) educating young children of both sexes and older girls;
(6) taking charge of the house;
(7) acting as emotional confidantes and shock absorbers.

The first six she sees as predating the money economy and arising 'from the bedrock of necessity — these things cannot be left undone if the human race is to survive and life is to be tolerable'. Ironically these essential tasks, and their institutionalized successors such as nursing, are consistently ascribed low social worth. In contrast 'men's work' can be seen as activity which originated in time and resources left over from the basic tasks. These resources could be diverted to coordination, trade and waging war. Novarra is not alone in suggesting that 'men's work' is a largely contrived and artificial activity. Dennis Pym (1980), for example, distinguishes work from employment. Work is those activities which generate goods and services. Employment is the set of conditions an individual fulfills to receive payment for their labour. Employment emerges as a set of rituals, many of which disguise the fact that no 'real' work is being done. Pym calls these rituals 'pseudo work'.

Some of employment's most fundamental rituals are those artificial criteria which set it apart from work — its hours of compulsory attendance, the centralized location of activities and the value placed on continuity of employment. These very rituals make it difficult for women to participate fully in employment without neglecting their basic tasks. The contract for this arrangement is sealed with payment of extrinsic rewards for employment itself, irrespective of 'work' done. Thus associations emerge between notions of work, status and sex which parallel those in my earlier analysis of language structures. Social value is ascribed to the work men do, and is defined by artificial barriers which prevent women's full participation. A critical example of women's exclusion from this system is the paradox that their experience of housework as 'work' is not expressed in established definitions of that word.

The current job market appears more complicated than a simple dichotomy between men and women's work would suggest. Novarra has, however, captured its deep structure of relative values. Even if employed, the vast majority of women are in low-status, low-skills, low-pay jobs. Chapter 4 returns to this topic with a more thorough look at what places there are in organizations for women.

Private and public worlds

I shall now draw on the man-made distinction between 'public' and 'private which appears repeatedly in the literature about women, to explore the world of women managers further.

The 'private' women's domain of the home is often distinguished in analyses from the 'public' world outside which is identified as men's arena. Even when women enter employment, many of their roles (such as secretary, assistant, personnel officer and tea lady) are so reminiscent of traditional female stereotypes, that they are essentially pockets of the personal, or private, within the bureaucratic, public, organization. Most of these traditional women's jobs offer few or no opportunities for career progression. They provide support rather than competition for male organization members.

One consistent factor, whatever line is drawn between public and private, is that 'Women's place is *outside* the ongoing action' (Miller, 1976). This corresponds to being outside the social construction of meaning in the above linguistic analysis. Women are always on the boundary's private side. As a consequence, any moves they make towards participating in public life violate fundamental social rules. Much of the relevant code has already been described.

Language and the nature of conversation render women silent and invisible. A woman speaking, or writing, in public therefore represents an inherent contradiction — both by crossing into the public sphere and by breaking the silence patriarchal order demands of women. The woman manager stating her case at a meeting crosses this boundary, and is likely to be sanctioned and reminded of her place for so doing. Many women in responsible jobs report that they are interrupted, ignored, patronized and ridiculed in such circumstances, but that men making similar points are listened to and taken seriously. There are also repercussions of the heightened visibility which accompanies women's public action. Mistakes become more apparent, and individual characteristics, often those such as dress which are unrelated to job competence, come under critical scrutiny. The woman herself may feel uncomfortable at being the focus of attention, as she is more used to the rules and strategies of invisibility. Women's supposed concerns with clothes and physical attractiveness serve partly to achieve an appropriate public mask which reduces the need for on-the-spot maintenance of a suitable performance.

The theoretical literature about how social interaction is maintained helps me take the above ideas further (for example, Goffman, 1959; Mangham, 1978). It draws many of its metaphors from the theatre to explain how social order and apparent cooperation is achieved *despite* individuals' differences and strong needs for independence. The distinction between personal needs and ambitions and social action, is yet another way of saying 'private' and 'public'. I shall suggest in Section III that men are generally more motivated towards independence and women towards cooperation as basic life strategies. Because they focus on individual independence the theories I draw on in this section may, then, reflect male-biased means of achieving social order. This qualification is worth remembering, but as male models dominate society for the time being, this is a wholly appropriate bias.

Writers about negotiated order identify public action as 'performance'. Achieving the appropriate image and appearance become critically important. Rules govern the shape of performances — these rules may be general principles such as stereotypes and social norms, or more situation-specific 'scripts'. Performance often requires (private) rehearsal and preparation. Women stepping onto the public stage as managers are at a disadvantage on several of these dimensions. Their appearance does not fit that traditionally expected of a manager. They may not know or understand some of the relevant rules of performance — especially those based in male socialization or fostered in all-male environments such as the rugby club. Whatever their social learning, they will find that stereotyped social scripts in which male is superior and female inferior will intrude into situations and influence their or other people's behaviour. Many women also do not have the supporting cast of helpers to prepare their performance on which men call, or use those they do have (such as secretaries or husbands) in radically different ways.

Above all, women may not recognize that there is a distinction between public and private to be observed. Throughout the management literature, women are accused of failing to appreciate the difference between the formal organization and its informal communication network, and that rewards are often determined by the latter (for example, Hennig and Jardim, 1978; Reif *et al.*, 1975). This apparent ignorance is indicative of a fundamentally different world view. For their part women often note, but refuse to observe, what they consider to be a false distinction between formal and informal organizational behaviour. They claim to be 'too blunt' for their own good because they will not engage in what they see as the unauthentic behaviour which demarcates this boundary. Many knowingly refuse to be falsely pleasant to people in high places despite organizational folklore that this is essential to assure career progression.

Women who do not adjust their behaviour as they enter the public stage run risks. Models of negotiated social order identify the public stage as 'real', the more important arena. There are implications that who one is there is 'the real test'. Everyday phrases such as 'he let himself down' and 'she did not do herself justice' testify to the primacy of public over private behaviour.

It is significant that the Women's Movement's key slogan seeks a women-based mode of action precisely by removing the artificial boundary between public and private. 'The personal is political' bridges the gap between these two worlds in both directions. It exhorts individuals to recognize that how they behave makes a public statement about their values, whether they intend it to or not. The slogan also reminds us that public male supremacy filters down to have an impact on *all* relationships between men and women to some degree; even intimate relationships cannot escape its influence. I am cautious about assigning meaning to 'public' and 'private' in this modern recombination. If the original distinction was false, adding its two parts back together again cannot provide an adequate description of unity. Some feminist writers use 'the personal is political' to advocate organized political activity. This is only one possible interpretation. Others, which pull free from the male, oppositional paradigms which swamp organized politics, are also evolving. I am still wondering what my own form of politics is. It seems to be coming out from silence to write a deliberately feminist book.

An essential complement to the notion of performance is that of 'audience' — a person or people who observe and judge our contribution. This concept has hovered in the background throughout this and the previous chapter. In their various roles of shaping and enforcing language, men are the main evaluators of women's behaviour, with the constitutional ability to reward or punish as appropriate. In Chapter 2 it was men who judged women unsuitable for management. Women's audience is predominantly male. Even its female members have been socialized into male values and tend to apply these as consistently as do their male counterparts. The criticisms of each other voiced by some women executives are indistinguishable from those men use. They both identify female colleagues as over-emotional, insufficiently dedicated, politically naive etc. Patriarchal order makes it difficult for women to develop, express and apply their own base of value judgements. Their early social learning that males are superior often prevents them even conceiving of this as a possibility. Men are established as the judges to whom women should direct their efforts and attention.

A major development of the Women's Movement is women taking other women as an audience which is no longer second best to men. This has made a radical difference to the areas of experience explored and the forms of communication used. Old boundaries are no longer relevant. In many ways the resulting new tide of women's literature maintains the ethos of the private and personal which has been women's traditional domain. This is twinned with a bolder, more assertive voice insistent on a public hearing. The growth in women writing for women, too, illustrates a merging of public with private which removed old demarcations between these two worlds.

Reflections on social values and women's roles

The above sections have depicted a pattern of female–male relations in which the former are consistently excluded from activities and spheres of influence which the latter have designated more real and more important than female domains. Continuing with the theatrical metaphor, it becomes appropriate to look again at what roles women *do* play, if they are restricted from equivalent parts to men in society's drama. Within organizations they have typically performed supporting functions such as secretary or personnel officer, 'outside the ongoing action' (Miller, 1976). They have, if invisibly, contributed significantly to organization functioning. In the more specific arena of conversational analysis, Fishman (1977) reaches parallel conclusions. Women expend a lot of energy sustaining conversation, doing its chores and keeping it pleasant. Fishman analysed many hours of taped conversations between mixed couples to find out who controlled the topic. She concluded that women made the conversational effort, but men determined what was talked about by whether or not they took up the topics which women offered in apparent attempts to draw the men out. Once a man did become interested, he would take over the conversation. Spender offers the following transcript from a social gathering as a near-perfect example of Fishman's conclusions. Throughout, male conversation is being encouraged at the expense of female:

FEMALE: Did he have the papers ready for you?
MALE: Mmm.
FEMALE: And were they all right . . . was anything missing?
MALE: Not that I could see.
FEMALE: Well that must have been a relief, anyway . . .
FEMALE: I suppose everything went well after that?
MALE: Almost.
FEMALE: Oh, was there something else?
MALE: Yes, actually.
FEMALE: It wasn't X . . . was it? . . . He didn't let you down again?
MALE: I'd say he did.
FEMALE: He really is irresponsible, you know, you should get . . .
MALE: I'm going to do something about it. It was just about the last straw today. How many times do you think that makes this week? . . .

Women's support provides an active endorsement of male supremacy which is apparently essential to its smooth operation. Spender concludes that 'It is not sufficient that males should be seen to be in control; females are required to be seen willingly supporting that control'. This account of a typical conversation epitomizes the asymmetrical pattern of relations between women and men described so far in this chapter. My next step is to probe more deeply into 'how' and 'why' such patterns maintain themselves despite their apparent inequalities. Why women support forms of relationships which appear to be so obviously to their detriment is a significant cultural mystery.

III: TOWARDS EXPLANATIONS OF SOCIAL POWER

Notions of social power are implicit in the framework of male dominance. It is the medium through which men shape and control society to reflect their experiences and serve their needs. Wanting to understand the forces which hold this pattern of inequality in place, I looked for suitable theoretical material. This section describes the two routes to explanation I followed. The first is a sociological view, identifying dominant and oppressed groups within society, which is popular amongst feminists as an explanation of women's devalued social position. I found this model illuminating, but only partly satisfactory, and moved on to a complementary analysis more concerned with individual psychology.

The synthesis of these two approaches eventually provides the framework I have been seeking within which to understand women and men, together or separately, *and free of values which devalue either sex*. This framework is the core for the rest of the book.

Anticipating the kind of conclusions this exploration might reach, I can identify two extreme possibilities. Women and men's asymmetrical relationship may indicate harmonious, complementary teamwork, or oppression by the latter of the former. The notion of teamwork has some advocates but its arguments are intrinsically flawed. They depict women as willing allies to men's authority over them. As women really like and benefit from being dominated by men, the claim runs, they have no right to complain. Such reasoning is itself proof that women are not mutual partners — because men can veto their expressions of concern.

Evidence favours the opposite viewpoint: that women support male dominance because they are so effectively socialized into male values, through cultural mechanisms which include language, that they are involuntary victims of a social system which consistently devalues them. From this perspective women are even cheated of any appreciation of their social position. They are therefore unable to develop their own frames of action. This explanation accords with much of the earlier discussion of language, and so gives me a starting point. As this section progresses, however, it becomes apparent that oppression is only part of women's story; there is more. However devalued their social position, women do have their own experiences and values, and an ability to act from them. This can be illustrated in relation to public performance. *In addition* to being excluded by men, women also have their own independent reluctance, based on notions of authenticity, about engaging in public activities. They may even derive a certain pride and satisfaction from their outsider status.

Seeing men and women as dominant and muted groups

Analysis in terms of broad social groupings has recently been popular amongst writers seeking to describe and explain relationships between men and women. These writers distinguish two social groups — one dominant and the other subdominant. The latter is often called 'muted' for reasons which will become clear. I hardly need to explain that men are identified as 'dominant', and depicted as using their overt power bases to maintain an essential inequality between the sexes.

I shall draw on three authors' descriptions for an understanding of the dominant–subdominant model as it can be applied to sex groups. Miller's analysis which appears in 'Toward a New Psychology of Women' (1976), was originally developed in this context. Spender (1980) is still a valuable source. For a third viewpoint I looked outside the field of women's studies. Freire's (1972) work is based in an examination of literacy as a potential force for social change in the Third World. I shall first briefly review the essential elements of this approach to explanation.

The categories of dominant or subdominant are ascribed to individuals by birth, and defined in the context of permanent inequality. Maintenance of these relative positions of power serves primary functions for the society or individuals concerned, and so acquires a force of necessity well in excess of its apparent usefulness. Any dominant group member championing the subdominant group's cause is therefore likely to be severely punished. The subdominant group is kept in place by being labelled 'substandard' in a variety of ways, and is ascribed appropriate social roles. Any 'good' characteristics particular members display — that is those which match dominant group's norms — are either ignored or dismissed by calling them abnormal. Placed in this position the subdominant group becomes preoccupied with basic survival. Their most sensible overall copying strategy is accommodation or adjustment. The subdominant group thus comes to accept and believe the other's definitions, including those evaluating the two groups' relative merits.

Through such social processes, one social group's meanings control the culture: 'Inherent in this analysis of dominant/muted groups is the assumption that women and men will generate different meanings; that is, that there is more than one perceptual order, but that only the 'perceptions' of the dominant group, with their inherently partial nature, are encoded and transmitted' (Spender, 1980). As their model (or models) dominate the social system, members of the dominant group act as chief evaluators and arbitrators of what is acceptable. They are the audience for subdominant group members, who have only limited powers to assign public worth from their social position. One repercussion, which helps in its turn to strengthen the system's basic operation, is that members of the muted group come to define their own aspirations in terms of dominant group goals. This undermines their identification with their own group's concerns, and reduces their desire and potential to challenge the dominant group. Some of women's current demands for equal opportunities are clearly borrowed from men's values about desirable lifestyles, and can be interpreted in these terms. As Callaway (1981) points out, however, if women seek 'liberation' by copying men they are still essentially acting from a muted consciousness.

To avoid negative sanctions, the subdominant group must be continually sensitive to the dominant group's needs, and play down its own. It therefore makes understanding the world around it a priority. Listening is the basic skill members develop to achieve this; it is their key to understanding others' views of reality. Spender suggests that women's so-called 'intuition' is often simply the result of listening rather than talking oneself. By developing strategies which maximize their understanding of what is happening around them, women achieve what Schowalter (1977) identifies as 'the best seat in the house' — the theatrical metaphor again.

The muted group use their information to sustain the dominant group's definitions of reality. They also develop attributes which please the dominant group, thus further ensuring their own survival. When it does need to do so, the subdominant group operates indirectly, and quietly, to avoid attention. It will tend to express itself through the other group's language — for example, the translation process of 'telling it slant'. Such behaviour earns the subdominant group the alternative designation 'muted'. Freire says that we find a subculture's 'themes' or core characteristics in the way the group thinks about and faces the world, particularly in the way it responds to critical sets of circumstances which he calls 'limit situations'. The subculture's themes and characteristics may not be expressed concretely, but this, in itself, is instructive: 'The theme of silence suggests a structure of muteness in the face of the overwhelming force of the limit situations' (Freire, 1972).

Because its expressions of experience are silenced or distorted the muted group cannot communicate fully amongst members. They are particularly restricted from making their meanings public, or from leaving 'traces' which can inform subsequent generations. An example for women is that expressions of the more painful, unrewarding aspects of childbirth and motherhood, which many women have experienced individually, are only recently penetrating the public media to qualify its otherwise idealized images. In another context, the United States' treatment of Vietnam war veterans shows similar processes at work. They are being socially ostracized and ignored by official care agencies because they are embarrassing reminders of an ignoble war. These, and similar examples, show that the muted group's concerns or qualities are seen as a threat to the dominant group's own needs and psychological well-being.

This perception of threat provides the underlying dynamic for the dominant/muted framework. The muted group is suppressed and held in place precisely because it represents a challenge to dominant group security. By taking containing action against muted group members, the dominant group is able to control its own anxiety at the images they represent. Feminist analysts explain that men assign the aspects of human functioning which they experience as threats — particularly emotions — to women. By devaluing women and designating them 'trivial', men master these aspects and are able to cope with their own anxiety. '*Woman* has become synonymous with chaos, but, by controlling women, the illusion of overall male control remains intact' (Spender, 1980). Oakley (1981): 'Women's instability stabilises the world'.

I have already described many of the mechanisms by which men maintain their authority in society. Language plays a particularly powerful part in suppressing the inherent conflict between women and men by keeping it outside consciousness. Lukes' (1980) 'three-dimensional view of power' identifies the inherently subversive processes at work. Earlier sociological theories of power had focused either on face-to-face exchanges in which one person imposes their will on another, or on an individual's or group's ability to shape decision-making to favour their outcomes. Lukes' further dimension of power illuminates 'the many ways in which *potential issues* are kept out of politics, whether through the operation of social forces and institutional practices or through individuals' decisions'. A person, or people, can so shape the world of others and influence their consciousness, that the latter fail to recognize an inherent contradiction between their own real interests and the interests of those exercising power. Lukes uses the notion of 'latent conflict' to capture these circumstances as 'there *would be* a conflict of wants or preferences between those exercising power and those subject to it, were the latter to become aware of their interests'. The implicit conflict remains latent, however, because it is submerged, and therefore unrecognized. Even if the conflict is vaguely sensed, it will prove difficult to pursue and develop, because of its departures from the taken-for-granted world of accepted concepts. Unorthodox opinions can be easily censored on this basis alone. Male society and science is based on notions of objectivity, for example. In their accusations that women's writing is 'subjective', commentators are using 'a ready-made format for dismissing feminist meanings' (Spender, 1980).

Dominant group members' practices of virtual censorship help to keep members of the muted group in ignorance of how widespread their concerns are, and so divided. Muted group members relate to the world of public communications either through dominant group gatekeepers, or through the latter's artefacts such as the published media. Ironically, the subdominant group has the wider base of information, one of the traditional sources of power, because of its pressing need to monitor, and respond appropriately to, its audience. The dominant group are typically deprived of the other's knowledge, even as it relates to themselves. (But this is often information they prefer not to know.) Thus the dominant group's view of the world, and potential development for, is impoverished by the very patterning they foster.

Reflections on the dominant–muted groups approach

An analysis of society in terms of dominant and muted groups has many inescapable parallels to accepted views of men and women's relative social positions and developed characteristics. I learnt more about language as a specific case from delving into its implications and interconnections. But I was still dissatisfied, sensing that this view of the world was somehow partial. I soon identified the causes of my irritation. Because of its sociological roots, the dominant–muted framework makes sweeping statements about individual behaviour that do not match my experience. I risk confusing logical levels if I try to make sense of people as individuals using sociologically valid theorizing. The approach also places too much emphasis on the public arena. On both counts, the framework leaves individual motivations largely unexplored.

I still know relatively little about women and their perceptions of their part in he total social system. I have learnt more, yet again, about men's perspectives. The enigma that women, who appear to be so disadvantaged by its operation, act largely to support the current pattern of relationships still remains. As Spender, puzzled, but hopeful, notes: 'It is paradoxical that part of the mechanism for ensuring the continued silence of women lies within women's control. They can cease to collude, they can abandon their role as willing slaves'.

Neither does the model help me understand close relationships between women and men. Winship (1978) suggests that there is a major tension in the lives of many women who do believe that they are socially disadvantaged but also want to form satisfying personal relationships with men. She concludes: 'Thus at one level women's oppression in the family is *experienced* and understood, its implications of subordination are finally denied, resolved only within the personal relationships of individuals.' In its application to other sociological contexts, such as white–black or government–peasant relations, the dominant–muted approach assumes relative social isolation between the groups for its dynamics to accord with theory. As several writers have pointed out, men and women deviate from the framework in their development of intimate relationships. As Farrell (1975) puts it: '*This is the only revolution in which the so-called oppressed is allegedly in love with and sharing the children of the oppressor*'.

Intimate contact between men and women acts partly to sustain the social system of male dominance. Woman's major social roles are defined by their relationships to men, limiting their allegiance to each other and the potential for joint action. Many women may not distinguish men as a separate social group, and so are unlikely to identify them as agents of oppression. The working through of conflicts within a relationship can also act as a private safety valve preventing latent social tension from reaching explosive proportions. Alternatively, or possibly simultaneously, intimate relationships may serve to moderate male domination. In their private arena women have an opportunity to establish an identity, skills and influence which are recognized and valued by both themselves and other family members. These can serve as a base for relations which contradict the dominant–muted framework's predictions. Certainly it seems unrealistic to analyse either social group in isolation. The subdominant group's experiences and models of reality are shaped *in part* as a counterpart to the primary communication system. Dominant groups' concepts and structures also incorporate an awareness of muted group themes.

My second major reservation about the dominant–muted framework is the primacy it gives the public world. The dominant group, men, are still the focus. Women are fitted into the gaps around them, and are given shape only in terms of their responses to male culture. I cannot accept this patterning. It implicitly, and fundamentally, devalues women by devaluing the muted position to which it assigns them. This devaluing of women comes from the model's assumptions of competitive power, but has filtered into analysts' own thinking. Once established it drives out alternative valuing perspectives on women's roles and characteristics. But I hold to my belief that there is more to women's world than their relationships to men. The muted tradition can be identified as muted without being devalued; it can instead be respected and honoured.

Once I recognized an implicit bias towards the public world in judgements about the muted tradition, I saw that other feminist analyses had taken similar directions. By paying attention mainly to the public, many feminist writers seem to have fallen into the trap of adopting a core dimension of men's definition of what is real, despite rejecting this definition as a whole. Through this selectivity, they often devalue women's own arenas, and so act to confirm rather than question male values. One repercussion is a concentration on the (male) world of paid employment at the expense of home, family, social and community relationships. This discrimination violates the Women's Movement's principle of developing positive notions about women based in women's own terms of reference. Instead it parallels, and therefore reinforces, the value judgements men make, and the status they accord women. Spender also comes close to reinforcing this bias in her analysis of language. She describes women as having 'negative semantic space' and being unable to develop positive meanings for themselves. In pointing to women's *public* silence she risks failing to 'hear' women's own statements about their worth based in their own forms of expression. These may take more symbolic forms such as myth or ritual, or be passed on through a private, one-to-one verbally-transmitted culture. By neglecting alternative modes of doing and being, the dominant–muted analysis and others which similarly take the public world as the more real, do not allow women to have values and experiences of their own, or to interpret the world independently of men's values.

The challenge to feminists is what values to adopt or retain when so many are based in patriarchal culture and priorities. Unless we develop perspectives on the world which do justice to the roles women have played, as well as to women's potential places in men's arenas, we lack a satisfactory base from which to choose. A first step is to listen to the 'voices' of the muted tradition. This involves not merely identifying its messages translated into other modes of speech, but appreciating their original nature, forms of expression, basic values and motivations. Several influential feminist writers (for example, Friedan 1982) are moving in this direction by advocating a renewed concern about women's traditional domains, to balance recent concentration on the public world.

Once I became clear about the dominant–muted framework's flaws, I could identify my next step. I looked for ideas which would extend the framework into individual's lives, and give equal worth to non-equivalent characteristics or social positions. At last I found I could write women into society as women rather than having to interpret them through their relationships to men.

Individual strategies for living: agency and communion

I found the fresh approach to explaining the pattern of male–female relationships I was seeking in a theory which has been used by other feminists to reflect on different styles of doing research (Reinhartz, 1981). It proposes two distinguishable strategies with which individuals respond to the core life issues of how to deal with threats in the environment, come to terms with death, and reconcile personal and community interests. Bakan (1966) calls these two fundamental tendencies, or principles of human functioning, 'agency' and 'communion'. The concepts take me away from sex differences for a while. I shall first describe the two basic principles and how they relate to each other, doing so at some length because of the importance these ideas had for my own thinking. Then I shall use them to make sense of men's dominance of social power, and draw implications for women managers.

Bakan proposes agency and communion as twin styles that individuals use to resolve core dilemmas of existence: those of 'being and not-being; and of independence versus interdependence. In essence they are basic coping strategies for dealing with the uncertainties and anxieties of being alive. Agency is the expression of independence through self-protection, self-assertion and self-expansion; communion the sense of being 'at one' with other organisms. The agentic strategy's main aim is to reduce tension by changing the world about it; communion seeks union and cooperation as its way of coming to terms with uncertainty. Whilst agency manifests itself in focus, closedness and separation, communion is characterized by contact, openness and union. The tendencies are potential complements rather than alternatives (but their very splitting conceptually is a product of the agency feature). The two different styles of handling uncertainty can best be explained further in hypothetic action. Table 3.1 summarizes their characteristics. I shall start with agency as the easier to describe.

Agency

This is the tendency associated with the ego, with the individual's transactions with the external world. Bakan identifies four processes through which agency accomplishes its core aim. These are based on the principle of achieving control by projecting difficulties one cannot cope with outside the self. The first process is *'separation'* or splitting. The individual distinguishes between what they 'like' and 'dislike', and represses the latter from consciousness. This often takes the form of projecting these qualities onto a person, group, object or concept in the environment which is then devalued because of its associations. In this sense the agentic uses 'knowledge' instrumentally, to act on the world around it. The individual thus reaches the second process of *'mastery'*. Paradoxically, this form of mastery is continually vulnerable as it leaves so much uncertainty beyond its 'control'. It is therefore supported by *'denial'* as the third essential agentic process. By denial the individual screens out perceptions of the environment and their own feelings which threaten the stability and mastery they appear, on the surface, to have achieved.

In these three stages I have already covered the usual explanation of how the dominant–muted division in society is maintained. Men 'agentically' project

Table 3.1 Summarizing the characteristics of communion and agency as life strategies

	Agency	Communion
Main aim	Control Independence	Union Interdependence
Dominant strategies	Assertiveness Control through: separation/splitting; projection; mastery; denial; encountering the projected threat Change	Cooperation Contact Openness Acceptance Personal adjustment
Characteristics	Doing Egoic Formal organization Physical action Classifies, and projects classifications onto environment Distance Contracts Change-resisting Achievement-oriented	Being Tolerance Trust Naturalistic perception of environment: emphasis on wholes, patterns, relationships, contexts Emotional tone Non-contractual cooperation; forgiveness Change-accepting Contextually motivated

threatening qualities onto women, and master these threats by suppressing and devaluing women. I am enthusiastic about the notions of communion and agency mainly because I can go beyond this interpretation, to gain more insight into the nature of agentic action, *and* explore an alternative mode of human functioning — that of communion.

The fourth stage in the agentic cycle is that of 'healing'. The individual comes to terms with what has been repressed by *'beholding'* or fully encountering it. Doing so is a precondition for reuniting what was originally split apart. This stage is essentially paradoxical. In order to accomplish it successfully the individual must suspend the initiating ego with its powerful need for control, and engage in free association with the originally repugnant element. This requires suspending belief and judgement, and temporarily setting aside anxiety. In relationships, for example, trust is offered and accepted. By these processes the individual substitutes understanding for control. 'Here is contained the paradox that the surrender of mastery in the ego's sense results in a more profound mastery' (Bakan, 1966).

Communion

This is not a modality of sequenced stages. It functions all at once, and as a continuous realm of possibility. Communion's main strategies for dealing with the world are acceptance and personal adjustment. Grounded in its orientation of union, communion's perception is naturalistic, reflecting the nature and patterns of the environment, and is only minimally guided by prior analytic classification. This provides a rich, multi-dimensional and dynamic appreciation of what is. As a result, communion's understanding tends to be in terms of wholes, of patterns of relationship, and of focal elements embedded in their appropriate context. The relationships between elements are more important than classification of the elements themselves. Communion's openness of contact with the environment produces intense personal impacts. These provide an extra dimension of sensing which contributes to understanding. As a result, emotional tone tends to be in the foreground, and the 'information' the individual seeks to communicate will include a significant strand of personal expression.

Whilst agency tries to stabilize its environment or make changes happen to its own dictates, communion accepts even noxious elements, and willingly adapts to circumstances. Its naturalistic understanding of the environment considers change natural. Although communion tends not to predict specifically, neither is it taken wholly by surprise. Communion incorporates an expectation of change — to the self, situation or context — as inevitable. Acceptance means that change is toned neither excessively positively or negatively — it just is.

In 'The Duality of Human Existence', Bakan explores the agentic and communion principles through science, religion and Freudian psychotherapy. Drawing extensively on myths and symbolic representations, he conjures for the reader concealed levels of associated meaning. I shall not present the details here, but will report some of the key associations Bakan makes through these analyses. They are surprisingly reminiscent of themes already introduced. He suggests, for example, that to avoid the anxieties involved in direct contact, the agentic substitutes formal social organization — the public world — for its potential private relationship with God. Agency emerges as a principle of doing, associated with the physical musculature, wanting to be judged against concrete achievement, entering into 'contracts' defined in fixed terms of time and money. These are the very conditions which distinguish work which constitutes employment from other forms of productive activity. In contrast, communion is better described as 'being'. Tolerance and trust are its characteristic manifestations. It works through non-contractual cooperation. Rather than extract retribution, parties show forgiveness, if either contravenes expectations.

Finding applications for the ideas of communion and agency in the world around me, I soon realized how widespread agentic strategies and values are. Science, even *social* science, is preoccupied with classification, prediction and control. Our image of a healthy, mature adult also advocates being in charge of one's life in these ways. Several writers, in fields from the academic to the popularist, are now advising us to redress this imbalance. In their various ways they are all calling for attention to the neglected alternative strategy of

68

communion, which may be more appropriate than agency to many aspects of a complex changing world.

The differences between agentic *doing* and communal *being* are important, and therefore worth drawing together before I progress further. Agency engages in idealization and tries to change the environment to match its own preconceived images. Doing is directed by internal, personal objectives. Communion is not inactivity in comparison, but its activity emerges from radically different roots — from its open contact with, and appreciation of, the environment. I have called this orientation 'being', it is mainly contextually motivated. Prior awareness and acceptance of the world as it is results in action which is in tune with the surrounding context, but is not conceptually premeditated. In a commentary to Wilhelm (1972), Jung captures this paradoxical essence as: 'The art of letting things happen, action through non-action'. Action based in communion may be highly appropriate as a result. It also risks being too thoroughly shaped and distorted by the environment.

Recent developments in sports training provide an illustration of communion's potential contribution to activity. A balance of communion with agency is advocated through the paradoxical state of 'relaxed concentration'. By 'quieting the mind' and 'letting things happen', this moderation of agency allows individuals' 'natural' abilities to operate unhindered by interference from cognitive processes. This requires a suspension of agency's judgemental function — the self who says 'You missed . . . people are watching . . . you'll have to do better than that!' Instead players are advised simply to pay attention to what is. In tennis (Gallway, 1979), this means the ball, the positioning of their racket and feet, where shots land on the court, etc. Through attending to being rather than doing, they free their abilities of coordination, and hit the ball more often and into more appropriate places.

The agentic mode interprets the world as its product because of the deliberate intent involved in its doing and the principle of independence in which this is based. From its perspective, 'success' is demonstrable and individual. In contrast, communion sees itself, and even its actions, as part of a wider context of interacting influences. It tends not to assume personal accomplishment when events turn out favourably.

Agency and communion in relationship, and their degenerative tendencies

Having mapped the two modalities separately, I can explore how they fit together, and each strategy's values and degenerative tendencies.

I initially proffered agency and communion as potentially equivalent styles. Their inherent qualities, however, shape their relationship towards non-equivalence and inequality. Agency's insistence on control has particularly profound effects. The distinction of communion from agency is itself a function of agency's attempt to suppress and deny the experiences of open contact with the environment which threaten its control. Agency therefore tries to repress the attributes of communion from which is has initially separated itself.

Through its own activities, then, agency creates around itself a world of competition within which it is 'naturally' the dominance-seeking style. In relation to agency, communion's cooperation-seeking strategies are effectively complementary, but fated to submission rather than equality. Acceptance of the environment becomes subjugation, rather than mutually influential union. In their pure forms, then, agency and communion can become locked in a spiral of potentially destructive effects. This analysis can be applied to the behaviour of social groups or individuals.

Degeneration of agency occurs if threats which have been suppressed to achieve control are not later recognized and re-incorporated. Instead, action on the environment is often interpreted as successful, and any conflicting evidence, such as unintended consequences or other people's disagreement, ignored. Repression of a communally oriented social group by one which favours agency stores up a legacy of resentment, which may eventually fuel a challenge to the dominant group's authority. Over-control can also unknowingly damage elements of the context or its patterning, whose importance were not originally appreciated in the narrow perspective taken. We are seeing this now ecologically, in the pollution of the environment and the expenditure of finite resources which are the by-products of meeting short-term goals. Eventually, the ultimate fragility of individuals or groups employing agentic strategies may be revealed, as they are crushed by forces outside their control.

Communion too has its degenerative cycle. Its strategies make it open to penetration, flooding and eventual destruction by contextual forces. The nature of its environment thus affects communion more than it does agency. This vulnerability is accentuated because communion has no strong base of self-worth from which to sustain itself — its attributions of 'success' are also context-dependent.

The two strategies do also have significant strengths. Each deals with uncertainty and anxiety in its own way, and can be highly successful in appropriate circumstances. Agency's main achievements come from thrusting out into the unknown, pitting its wits against environmental forces, imposing organization on chaos. Exciting engineering achievements are prime examples, as are space programmes. Communion's triumphs are, but by nature, less tangible. They come from integrating, reconciling, synthesizing, and from supporting the flow of events.

Each strategy if used alone continually walks the tight-rope between the possibilities of success and degeneration. Either can become an over-determined pattern, robbing the individual or social group of flexible strategies for coping. An attendant risk is that of energy mis-spent. It may require more energy to sustain an inappropriate strategy than it would to deal appropriately with the original threat. Use of the two styles in combination or synthesis offers a broader base of potential coping. A concern about appropriately combining communion with agency echoes through Bakan's book. Starting firmly in the agentic, he favours 'mitigating agency with communion' as a life strategy. A parallel notion originating in communion as a dominant strategy might be 'communion enhanced by agency'. I shall return to this suggestion below. Business meetings provide one opportunity to observe agency and communion as two modalities of human functioning in practice. Many meetings open with attempts at 'premature agency'. Someone is keen to get things going and takes the lead, making lots of suggestions. These are either openly or obliquely rejected. Other people are not ready, but the self-elected 'leader' has not recognized that yet. Discussion flounders; members generate apparently trivial or irrelevant ideas and objections which reveal much less fundamental agreement than the original prime mover had implied. Eventually, if the meeting is lucky, someone who has previously said very little (but has been listening), makes a suggestion which seems to integrate and satisfy different peoples' concerns. This is 'salvation through communion'. Meetings seem less prone to the alternative pattern in which individuals devote so much energy to understanding each other's points of view, that they fail to acknowledge fundamental conflicts of opinion, or to agree appropriate action.

The framework also helps to illuminate professional roles such as those of trainer and organizational change agent in which I become involved. Both involve dilemmas about the balance of responsibility between the professional and the client. The professional who uses a dominant approach of control will foster client dependence, and possibly even antagonism. At the other extreme, communal strategies may be uncomfortable because they contradict traditional expectations of professional expertness, and provide little structure to contain both parties' anxiety.

I can also identify communion and agency strands in the process of writing. Neither agency or communion on its own will get this manuscript intelligible and to the publisher.

Communion, agency and sexuality

The concepts of communion and agency helped me to explore further potential differences between women and men. At the level of aggregates or groups, communion can be broadly identified as a female and agency as a male principle. The two styles address fundamental human needs, and I expect women and men, from their different anatomies, socialization experiences and learnt, well-practised, social roles, to view these issues differently. As a result, I expect each sex to favour, although not be exclusively limited to, one set of coping strategies rather than the other. Anatomy provides the least socially contaminated perspective on potential agency–communion differences between the sexes.

Male sexual characteristics of thrusting, penetration, firmness and activity have clear links to the agentic. It also seems that males are more vulnerable to environmental influences than are females, and therefore have a more pressing need to control them. Statistics show greater male than female mortality in every age group from prenatal onwards (Bakan, 1966). Some of these causes of death can be interpreted as a result of a more agentic approach to life. This complicates the overall picture, but strengthens rather than weakens the association between male and agency proposed. Male 'vulnerability' may operate at a more micro level too. Research studies suggest that physical homeostatic mechanisms — such as body temperature, basal metabolism and blood sugar level — operate within limits narrower in men than in women (Terman and Tyler, 1954). This material implies that women can make greater physiological adaptations to changing environments, whilst men achieve necessary adjustments by modifying the environment instead.

There are evident connections between communion and female sexual characteristics of opening up, receptivity, and experiencing radical physical change through the monthly cycle of menstruation and through conception and pregnancy. Women's apparently wider range of physiological adaptation in relation to changing environmental conditions adds to this profile.

These sex-linked profiles are no more, but no less, than potential predispositions which act as the grounding for individuals' later development. I shall next use them here to introduce three tentative lines of enquiry which connect agency and communion more closely with women and employment, and return to their interplay in individuals' personality development in section V.

Some implications of identifying communion as a female disposition

Issues of *personal identity* figure significantly in later chapters. Distinguishing communion and agency as sex-linked strategies of human functioning sheds new light on popular concepts about identity and sex differences. Social norms idealize a strong sense of self-identity, self-esteem and confidence, based on a personal, independent assessment of one's own worth. These evaluations have strongly influenced social definitions of 'healthy' human development. I have noted an association of male with agency, and of agency with the independence pole of the independence–interdependence polarity. Seeing the individual as the most meaningful unit of analysis is itself an illustration of agency's tendency towards independence and conceptual splitting. Concepts of identity based in independence are therefore revealed as agentically based, and thus male-grounded. Women are often said, in contrast, to have 'relational' identities (for example, Gilligan, 1982). This paradoxical phrase means that women see themselves more in terms of their relationships to others than via a strong sense of self. It tends to have negative associations with 'dependence' and therefore 'weakness'. Viewing this alternative as an expression of communion, the interdependent principle, gives it a newly positive tone.

A significant issue submerged under the independent–relational identity distinction concerns who benefits. This brings me back to the two modalities' degenerative tendencies. Communal energy is expended for others, and possibly for oneself as part of the whole. Agentic energy is devoted primarily to self. For an individual, or group, to sustain itself successfully and maintain a viable environment, it needs to define itself in both self- and other-valuing terms.

So far I have depicted several ways in which *agency and communion translate into individuals' lives*. To explore this area further I need to move beyond the generalization that men are predisposed towards agency and women towards communion, to more speculative possibilities. The conceptual distinction between communion and agency is itself a reflection of the agentic feature. This offers a possible model of male as a differentiated organism in which one part has been actively repressed. From this viewpoint female can be interpreted as communion and agency in an unsplit wholeness. The essential nature of female thus becomes even more difficult to capture appropriately especially in analytic, distinction-making language. This would certainly not mean a simple sum of the two parts, which are anyway artifically distinguished categories.

Several sources suggest that the female principle is essentially dual in nature. Bakan interprets anatomical evidence to mean that 'the female of the species is more conspicuously genitally androgynic than the male'. Socialization theories reach similar conclusions. Chodorow (1971), Spender and others claim that society is far more strict in trying to ensure that little boys grow up male than that little girls grow up female. Calling the latter 'tomboys' raises less anxiety than calling the former 'sissy', for example. In their development of non-verbal behaviour, too, boys learn a stricter code than do girls. A major task of socialization is moving from close (touching, holding) to distanced (looking at, speech) modes of social interaction. Research data suggests that boys move faster that girls to distanced forms of relationship, and that girls are never socialized as thoroughly in this sense (Lewis, 1972). These illustrations suggest, in fact, that little boys are initially equally flexible in sexual disposition as girls, but have to achieve more complex and risky learning to be capable of manhood's separation and repression.

It seems likely, then, that women have free access to both communal and agentic strategies of being, and may be more flexible in this sense than are men. The aspirations many women currently express to have satisfying lives through both motherhood and challenging paid employment reflect a range of motivations which can also be interpreted in these terms. These women's objectives reveal both relational and independence needs, and are portrayed by them as elements in a total picture. Their superordinate goal is wholeness rather than specialized development.

Drawing these various strands together suggests a model in which communion is women's dominant tendency, but is twinned with a more or less fully developed agentic auxiliary. This combination is more viable than the alternative pairing of dominant agency with auxiliary communion. Agency will tend to repress alternative modes to achieve control, making this an inherently unstable pattern. Bakan's analysis emerges largely from a concern about this instability. If it is soundly based in an appreciation of its own value, I would expect communion to be tolerant and accepting of supplementary tactics for coping. There are several functions agency can serve as communion's auxiliary. It can provide protection for communion's vulnerability in hostile or new environments; create structures within which communion can operate; support communion by giving it direction; afford a systematic understanding of alternatives within which communion can locate itself; and help translate communion into direct effects through judicial instrumental action. My image for a positive direction for women, based in who they have been and are, is of communion enhanced, supplemented, protected, supported, aided, focused, and armed, with agency.

My third line of tentative enquiry takes agency and communion as a framework for understanding the current state of *female–male relationships in society*. Concern about women's rights and women's demands for equal opportunities have sent ripples through many people's lives. Without delving too deeply, the implications of identifying agency as a dominant male reaction and communion as a dominant female reaction, have a certain face validity.

I would expect men to be threatened by the surfacing of 'women's' issues which are reflections of their own denied aspects. Miller (1976) uses psychoanalytic evidence to confirm the nature and scope of this repressive activity.

'Psychoanalysis has in a very large sense been engaged in bringing about the acknowledgement of these crucial realms of human experience. It has done this, I think, without recognising that these areas of experience may have been kept out of people's conscious awareness by virtue of being so heavily dissociated from men and so heavily associated with women'.

She goes on to conclude that: 'Women are constantly confronting men with men's unresolved problems and challenging them with men's own unrealized potential'. If men deploy agentic strategies to counter these threats, they attempt to control, deny or devalue both the issues and the ways in which they are expressed. The double-binds they have so far offered women in the slogan 'equal opportunities' do indeed proscribe any questioning of fundamental values, and seek to constrain rather than free what women represent.

I would expect women's ways of reconciling (rather than 'solving' because that is not their concept), a new consciousness of men's domination of society with the conditions of their own lives, to be development of a quiet personal understanding by which to live in their private arena. Communal strategies are likely to find expression in a new sense of personal being. Outward signs might be a change in self-confidence and self-esteem, more clarity and assertiveness in relationships, and less, or no, guilt when 'failing' to match false ideals of femininity. Many writers report that this private shift in attitudes is happening, but opinions about the strategy vary. Some feminists bemoan this individualism because ideas and experience are not shared with those in similar positions and so cannot form a common basis for consciousness and potential action. Other writers suggest that whatever their mix of strategies, women currently have the edge in courage and thus potential benefits: 'Women are thinking out their roles; men are merely clinging desperately to theirs, hoping that they will survive the coming storm, searching for the means to prevent it happening' (Korda, 1975).

Bakan's suggestion to men would be the paradoxical fourth healing stage: relinquishing control to achieve greater mastery of what really threatens them. This can be interpreted in terms of agency's original mission: protection from the anxiety of close contact with potentially threatening forces. Men have, therefore, temporarily to relinquish control, suspending stereotyped perceptions of the world and the judgements which accompany them, in order to perceive clearly, and encounter, what is. Only by mitigating agency by communion in this attempt can the individual tolerate the attendant anxiety.

Merging an appreciation of communion and agency with the framework of dominant and muted social groups

I can now bring together my two approaches to explaining unequal power relations between women and men. The concepts of communion and agency, particularly the former, compensate for the deficiencies of the dominant–muted groups analysis without radically challenging its broad framework. My resulting model is a hard layer of the dominant–muted model on the outside, with a softer centre of theories about individuals and their different strategies for dealing with anxiety inside. The model recognizes male domination of women and their exclusion from the public stage; *but at the same time* values the muted tradition and its purposes.

Women are no longer *only* the projection of men's fears, and the support cast for men's competitive public activity, finding their only expression in muted group versions of male aspirations. Women have their own drama, too, based in fundamental strategies of individual and collective survival. Part of women's contribution to society is a widely attuned altruism, expressive of communion. Women maintain society through physical and emotional nurturing, often at the expense of their personal needs for independence. In doing so they tap a base of truth which values interdependence, and respects personal skills for achieving it. At an individual level women may also benefit by being helped to keep in check communion's degenerative tendencies to become emotionally flooded, by locating themselves within a framework of control.

The muteness of women's traditional activities should not affect estimations of their worth. It is not a sufficient reason for devaluing them. Doing so judges them by dominant group values. To understand women and their place in society fully, we need models which do justice to the pattern of relationships in private domains such as the family which have traditionally been theirs, rather than focusing on the public arena alone. A precondition of such models is that they appropriately value the communal, muted tradition and the experiences which express its themes.

Bakan offers one illustration of the directions theories based in cooperation rather than competition between women and men might take, in his model of sexual relations. His example seems a little idealistic to me, especially given society's persistent devaluing of the female principle; but it also has the hint of appeal and possibility:

'it is largely the agentic in the male and the communal in the female which bring them together. However, in the contact between the sexes over time, there is the cultivation of the integration of agency and communion within the male and female, corresponding to the integration of agency and communion between them . . . . This is one meaning which we can assign to the religious emphasis on the notion of "wholeness".'

Reflection

In the end, with its focus clearly on agency, Bakan's is a male analysis. His concern is how one can mitigate agency with communion. Bakan achieves this balance in his own work through a strong thread of myth and symbolism which is the book's testimony to communion. Women managers are probably more concerned to understand life when communion is one's basic strategy (and when others are keenly invested in you continuing to use it). I have suggested that 'communion enhanced by agency' moves in this direction. The following section continues with this theme. In it, I review women's resources. Much of this chapter has been deliberate or unconscious ground work for this task. Also, if communion has different ways of representing and reacting to the world, it is here I should look to find them.

To the task I take both new concepts, and a new sense of personal freedom arrived at during their exploration. The latter found three expressions, at least, of a very concrete nature in my own life: a new — faster and brighter — car; a pre-Raphaelite hairstyle and learning to fly. Having become more appreciative of my own communal and agentic aspects, I was able to experience and live each more clearly and fully.

IV: WOMEN'S CHARACTERISTICS AND CULTURE

Much of the chapter so far has involved comparisons between men and women, albeit with the pressing aims of disentangling the two, or distinguishing actual from idealized characteristics. It is now time to consider women separately. Achieving an adequate description of women has been a priority for many feminist authors. I shall draw on their work here. They have taken a variety of routes to achieve their pictures of womanhood.

Some analysts start from a critique of the male notion of femininity (see Chapter 2), and develop concepts based on women's own interpretations of their experiences (for example, Daly, 1979). Others take women's traditional sex roles and re-cast their previously devalued characteristics as positive (for example, Miller, 1976). Some commentators set out to rediscover a temporarily lost heritage, and look to myth and ritual for their sources. Claremont de Castillejo (1973), for example, suggests that 'Today women have become so immersed in a masculine world of ideas and principles that they forget their own basic truths'. Another group of authors has selected academic subject areas such as psychology, sociology and psychiatry which have been derived largely by and about men, and they have constructed the outline of a new discipline based on female developmental phases, experiences and meanings (Eichenbaum and Orbach, 1982). Finally, some feminist academics are developing new approaches to research. These tap female principles to arrive at humanistic, rather than agentic, models and methods of enquiry (Reinharz, 1981).

Whatever their approach, these writers are all involved in the same endeavour — the reappraisal of women's characteristics, social roles and culture, independent of male value systems. Callaway (1981) calls this process 're-vision', and identifies three ways in which this title is appropriate to women's current activities. She uses

' "revision" in the standard sense of correcting or completing the record; then "re-vision" as looking again, a deliberate critical act to see through the stereotypes of our society as these are taken for granted in daily life and deeply embedded in academic tradition; and finally, "re-vision" in its extended sense as the imaginative power of sighting possibilities and thus helping to bring about what is not (or not *yet*) visible, a new ordering of human relations'.

Daly uses the parallel notion of 'spinning' to refer to the process of creating new female-based meanings. Spinning has no one static meaning; its interpretations are as multi-dimensional and ever-changing as the creative process it designates. Spinning involves becoming free of male meanings, and embarking on the journey of exploring and articulating women's own concepts. Through this 'whirling movement of creation . . . spinsters can find our way back to reality by destroying the false perceptions of it inflicted on us by the language' (Daly, 1979). There are no constraints on the nature of women's journey — its only guiding principle is their sense of their own being and wholeness. 'Spinning involves spontaneous movement, the free creativity that springs from integrity of be-ing. . . . Spinsters spin and weave, mending and creating unity of consciousness'.

From their new insights and perspectives, women derive a new way of knowing which Daly calls 'lucid celebration'. She defines it as 'the free play of intuition in our own space, giving rise to thinking that is vigorous, informed, multi-dimensional, independent, creative, tough'. Daly's own writing is all of these things. Her use of the term 'spinning' is a prime example of her meanings. She retrieves the word 'spinster' from its current derogatory status, and from the patriarchal order which put it there. Re-investing it with its original meaning — 'a woman whose occupation is to spin' — she puts it to work *for women*. Daly says 'There is no reason to limit the meaning of this rich and cosmic verb. A woman whose occupation is to spin participates in the whirling movement of creation'.

Through the processes and experiences of re-vision and spinning, women can achieve fresh perspectives and groundings from which to understand them-

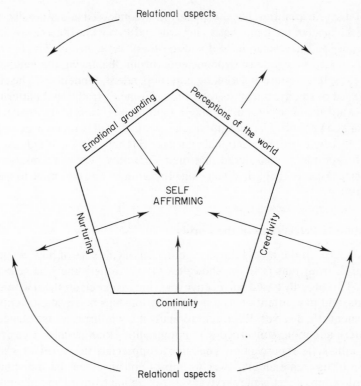

Fig. 3.1   Five dimensions of womanhood

selves and their relationships with men. They can then achieve real choice
based on an appreciation of options, and freed from the equally limiting
strategies of copying or rejecting. It is particularly tempting to reject in their
entirety women's previous social roles, once we identify them as the result of
oppressive socialization. Doing so is, however, over-hasty and largely defensive. I suggest we pause instead and spin a new base from which to make more
discriminating, women-centred, choices. These will involve some rejection, but
also re-vision. The woman-hours, experience, and skills women's devalued
social roles represent are vital elements in our heritage of resources, and the
base for our values. Valuing women means appropriately honouring this base,
and the spirit with which women engaged in it. Rather than renounce these
as part of a soured history, we need to reinvest many with new meanings.
Wherever we want to go next, we start from here.

What I offer in this section is another contributory step in the process of
re-vision. It takes the shape of a profile of female traits, brought together from
the various sources listed above, and drawing on the previous section's discussion of communion. The profile's main purpose is to map the many resources,
both communal and agentic, on which women can draw. It has taken provisional shape for me as the five-sided figure shown in Figure 3.1. Its dimensions

are not discrete qualities, but thoroughly interconnected and mutually signifi-cant. Each facet of the figure has an inward- and an outward-facing aspect— its significance to the woman herself and to others, respectively. The figure, the person, is seen in a state of dynamic equilibrium, balancing the independent forces for self with outward-looking, interdependent, tendencies. The state of balance is not steady, but shifts over time. Women's traditional pattern is for the outward-facing aspects in the figure to dominate. Recent data suggest this has changed very little. A questionnaire survey of 1500 women aged between 25 and 45, carried out by a national magazine (Hallett, 1982), found that only a fifth of respondents considered that their 'first duty' was to themselves. The self-facing aspects of each dimension, then, have the potential to become self-*ef*facing.

Dimension 1: Perceptions of the world

The first facet of the model draws together characteristics of how women see the world, from that position Showalter (1977) called 'the best seat in the house'. This broadly based perspective on events appreciates them within their contexts, and pays attention as much to relationships between elements as to the elements themselves. Representationally it is wholistic, aware of meaning-ful pattern and more concerned with potentialities than details. A basic moti-vator of this style of perception is needing to appreciate the world for what it is, because of its tremendous power to impact and threaten the individual. Its outward-facing aspect is the survival value of acting from an understanding of other people's viewpoints and emotional needs.

In a conversational setting this approach is demonstrated in women's skills in listening, but these are only one strand to a broader armoury of 'silence skills'. Such skills are sometimes dismissed as 'passive' because they contradict the strong social norm that doing is good, but passivity is highly appropriate in certain circumstances. Listening is one silence skill which has recently received some revised attention, however, mainly for its benefits in non-directive coun-selling. Interestingly, to show that it is 'good' in this setting it is labelled '*acitve* listening'. In this rehabilitation, the benefits so far identified are those for the person who is listened to. The values of good listening to the listener are still under-valued. Its risks are more frequently identified, and they too are illumi-nating. Listening involves opening oneself up to other people's views. Fully appreciating the complexities of this wider world limits an individual's ability to hold constant simplified opinions themselves. For those who value persistent and clearly oriented doing this kind of awareness would be a disadvantage.

This dominant perceptual approach also has implications for learning styles and for action. It suggests a style based on immersion, 'losing oneself', as a means of developing concepts and experimenting with potential meanings. It offers a grounded appreciation of contextual forces as a basis for action. This will tend to be translated into communion-guided influence, which works with and through its environment, rather than agentic control. Women's 'actions' and contributions to events may therefore be difficult to identify, as they seldom take the shape of discrete initiatives with clear personal ownership of results.

This perceptual viewpoint's hallmark is a full appreciation of the world as it is, through a sensitivity to intrinsic structures and qualitative patterning. Expressing what they therefore see, women have sometimes been said to 'demystify' organizational life because they interpret it without the learned concepts and norms men use.

Dimension 2: Emotional grounding

The clue to how these perceptions of the world are interpreted and understood comes from the second dimension — that of women's close relationship to emotions. I shall first consider some of its inward faces. As a result of their perceptual openness, women tend to experience the world initially through their feelings, rather than through predetermined concepts and classifications. In this way they are their own sensing devices, using an appreciation of the changes they experience in relation to the world as valuable data about that world. This sense-making strategy is often referred to as 'personalizing'. It generates particularly sound models from which to approach the world, because they recognize one's own and other people's emotions as significant aspects of the situation.

Keller (1980) has captured many of the emotional grounding dimension's themes in her profile of a female cognitive style. In doing so she renders the undifferentiated term 'intuition', with which these skills are usually (dismissively?) labelled, redundant. Even mentioning a female cognitive style is expected to invoke automatic negative reactions and so risks being an apologetic activity. Reinharz (1981), for example, asks her readers to suspend thoughts of 'sentimental, irrational, or unscientific'. Keller's description is, in fact, unerringly positively toned. Its main elements are: 'artistic, sensitive, integrated, deep, inter-subjective, empathic, associative, affective, open, personalized, aesthetic, receptive'.

This second dimension's *outward* face acknowledges the virtually commonplace observation that women are often facilitators to others' emotional expression and development. Referring to this role as that of an 'emotional shock absorber', Novarra implies that it is based in misplaced altruism. This interpretation is appropriate if women consistently suppress their own emotional needs whilst paying attention to other people's, but is a partial view only. Claremont de Castillejo (1973) interprets these activities more positively, identifying 'mediator' as one of women's most significant roles. In her broad definition she casts women firstly as mediators between society and the basic life forces. Their roles as child-bearers and raisers, and as nurses of the sick and dying exemplify this intercession. She thinks women also operate at a more individual level to help men contact their own unconscious forces and creative inspirations. This notion parallels the facilitation of agency by communion suggested above.

Again, we find women taking on roles which are incompatible with concepts of individual action and ownership of achievement. The contribution of mediators can easily go unnoticed. In an organizational context, such behaviour will not be readily identified as successful performance deserving career progression. Critical issues for women managers are what criteria *they* use to evaluate their own behaviour, and whether social values will change sufficiently to recognize their less visible contributions to organizational life.

Drawing on emotional grounding's understandings, women may also access alternative, female, modes of communication. Reinharz, for example, suggests that writing based on personalized research methods should seek an encounter with the reader by being readable, evocative, communicative and usable.

Dimension 3: Nurturing

The third facet of my figure uses the broad term 'nurturing' to identify the wide range of activities in which women engage to service the basic life-maintaining needs of society generally, and of their families in particular. This covers all archetypal women's work: child-bearing, care of home, husband, children and the sick; food preparation; provision of clothing and emotional support. This dimension's outward aspect, its value to others, is more obviously apparent than its value to the individual concerned. It does, however, have both instrumental aspects concerned with achieving identifiable goals and emotional aspects expressing the maintenance of relationships. The latter are its inward face. In this duality, though, also lies the dimension's chief potential for mischief. Confusing the two aspects provides fertile ground for misunderstanding and conflict. If a particular task, such as cooking a meal, is seen by its perpetrator as expressive of their relationship with others, but by the recipients as purely an instrumental means to an end, the two parties will be working from different, potentially conflicting, expectations. Demanding service functions as proof of an expressive bond is a similar confusion of these two levels of relationship.

Women are particularly vulnerable in their nurturing roles to exploitation, frustration of their own needs, boredom owing to the repetitive nature of many of the tasks, and guilt if they do not live up to their ideals which place these activities at the heart of a fundamental belief in family and community interdependence. In the notion of nurturing, the relational facets of women's identities become particularly clear. Self-worth is derived through collaboration with others.

Dimension 4: Continuity

Several of the above notions of femaleness have been underpinned by a sense of the importance of continuity — of enduring relationships over time, for example. To highlight what I see as a vital element, I have separated this time dimension out as a further facet of the model. Its outward face is towards maintaining relationships and continuing life maintaining tasks, whether or not these are satisfying at the time. Their continuation involves a sense of wider purpose than the particular moment. Looking inward the theme has strong connotations of change. This is most apparent at the physiological level, but is part of a more general pattern. In this cycle current state is important in its time, but it is essentially transitory. Life is a process of being *and* becoming; new forms of integrity are repeatedly achieved and relinquished. These themes are illustrated in managers' approaches to their work lives reported in Chapter 6. They want jobs which satisfy, enhance and develop them as people. They expect and want to be changed by the work they do. Their demands of the particular job as a position in the organization hierarchy are static in comparison. The managers look for current challenge, and most do not pursue a rational strategy of career which might sacrifice this for possible rewards in the future.

Key themes contributing to the continuity facet of women's profile are patience, resilience, endurance and personal transformation.

Dimension 5: Creativity

My final, fifth, facet of core femaleness is one I can hardly put words to, but know must be included somehow. It is the wild spirit of womanhood. Daly (1979) captures this most evocatively, and it shows through in other women's writing and being. This dimension's foundations are in creativity, and the untamed knowing which survives being muted and devalued. The concepts and language Daly invokes are associated with witches — 'the Great Hags of our hidden history'. She sees this tradition as suppressed by patriarchy because of the untamed force of womanpower it represents. In her writing she strives both to expose the negative labelling of these characteristics and to spin new positive meanings.

The inward face of this dimension is wild, intractable, untamed knowing; capable of deep-rooted laughter; strong and wise. Its outward face shows traces of pain and cynicism at society's blindness to interdependent, ecological concerns. It may appear threatening to others who appreciate the potential force of its energy, but fear that force is uncontrollable. Daly interprets the cultural mechanisms by which women are socialized to accept male authority, as expressing men's need to contain and master women's power. The creative dimension supplies the energy for female being. If it is suppressed or constrained women's sense of their own truths and values becomes numbed. The Women's Movement derives much of its radical fire from this source.

Relational identity: personal power

My profile of womanhood provides a framework through which to return to the notion of 'identity' as it has meaning for women. Rather than contrast relational identity, with which women are labelled, with its more socially approved independent alternative, I see these as complementary forces, balancing and mutually enhancing each other. The model's outward faces express tendencies towards interdependence which make powerful sense at a wider community and ecological level of survival. They are, however, vulnerable to others' disapproval (or indifference), and their accomplishments, by their very nature, are unlikely to contribute to a clear and lasting sense of self-worth. The individual needs to develop *as well*, factors which foster its own stability and sense of integrity. I have therefore set tendencies toward interdepence in a framework of balance, poise, dynamic equilibrium with forces towards independence. Paradoxically, these personal anchors will enhance individuals' abilities to promote interdependence.

Translated into practice for women, this model means pursuing two lines of development concurrently. There is no value in rejecting one partial notion — that of relational identity — only to adopt another equally partial — that of independence. Instead, women have the opportunity to develop a new sense of self which draws on, illuminates and enhances the values they already express in their mode of being, and takes these as the bases for self-worth. There are two keys to achieving this new, female-based, meaning to 'sense of self'. The first is establishing a base of values — both personal and collective — from which women can accept and respect themselves without relying on others' approval. This need not involve completely ignoring others' value systems. Feminists have not given up their relational identities, but have diverted their energies to dialogue with other women, rather than continue to 'receive' values from men. The second key is releasing the muted energy of the female characteristics described above. This requires recognizing the processes of muteness, developing a new awareness of oneself as a woman, and cutting free from the pervasive swamp of male norms and values. Only through such processes can this base of women's experience be free to express itself fully, and in its own tongues.

Women's culture

So far the analysis of women's characteristics has taken the individual woman as its focus. Another resource at which I have sometimes hinted is what may loosely be termed 'women's culture'. This a difficult area to explore and capture for its apparent contradictions. From certain perspectives the female infrastructure in society seems weak, fragile and insubstantial. Women's primary roles, even those within business organizations, are typically defined in relation to men who mediate between them and more public arenas. These relationships keep women relatively isolated from each other, particularly in terms of allegiance. Through their dominant survival strategies, women develop personalistic rather than collective solutions to this structuring. They tend not to share their current understandings in a wider community, and also do not have the benefits of cumulative records from previous generations' experiences.

These notions, however, tap only one aspect of female culture — its public, observable face. There are other facets of woman-to-woman communications his orientation fails to capture. As I have found before, seeking to express alternative perspectives becomes problematic, almost inappropriate, because of their largely non-linguistic (or a-linguistic) nature. From this alternative viewpoint what people say to each other fades in significance relative to the messages' expressed by being together. These messages are about emotional sympathy, tolerance of differences, solidarity, leaving space for others to be themselves and appreciating a space in which to relax, amongst other things. They do not rely on the passing of information for its own sake. Exchanging ideas seems somehow redundant to the exchange's underlying functions. More typical forms of expression are an attentive presence with and for each other, working side by side on a common task, and the expression of emotions. The theme of continuity is also expressed through the role many women play in discovering, remembering and passing on family histories about personalities and relationships. These activities have archetypal connotations, given the themes of connectedness developed throughout this section.

From a more academic approach, Bernard (1981) paints a similar picture. She proposes a two-dimensional characterization of 'The Female World'. Its structure, she says, is based in local, kin relationships — a legacy of pre-industrial society. Culturally its ethos is of love (which has erotic, philanthropic and humanitarian aspects) and/or duty. Men's world, in contrast, follows the model of economic, exchange-based capitalism.

In a recent seminar on women and men at work, I was a member of an all-female group which showed many of the qualities I have listed above as female qualities, and which to me was an enlivening place to be. It was a forum for feelings rather than ideas. The atmosphere was generally friendly and accepting. Differences of opinion were sometimes hinted at but not fully pursued. My main learning from this experience which is relevant here is that it is easy not to see what is happening in women's culture if we look through male paradigms of values. An authoritative observer interpreted the lack of open conflict as a sign of 'avoidance behaviour' and 'arrested development'. Their schema of what was normal and acceptable had no language for what the group had been doing, how it had gone about it or what its potential development might be. I was disappointed, but enlightened.

Afterword

In this section on women's characteristics and culture, I have tried to see women through their own eyes, helped by various writers engaged in re-vision. The emerging profile of womanhood is a cautious step in this woman-centred process. In the sequence of this book it also provides a framework through which I can start to make sense of the women managers' accounts of their lives which make up Chapters 6–8. There is now one further avenue of feminism to explore, before I draw this chapter to a close.

V: ANDROGYNY, AND ADDING SHAPE TO COMMUNION'S COMBINATIONS WITH AGENCY

As a final element in this chapter I shall consider the concept of 'androgyny' which has become popular recently as a way of reconciling the female and male principles. In my early reading I had found this notion both appealing and disappointing, liberating and constraining. Only once I had arrived at my own understanding of communion and agency, and of communion enhanced by agency as a positive direction for women, could I make personal sense of androgyny and identify its potential contribution to my thinking. It is a rich concept, full of potential; but it should not, I think, be pursued to a firm definition, at least for the time being.

In the literature on androgyny as a whole, two broad traditions can be identified. One emerges from myth and symbolism, often via Jungian psychology, and portrays androgyny wholistically, mystically, as a realm of possibility. This approach to the concept is through communion as a process of sense-making. At the other extreme, androgyny is dissected into parts, categorized, defined and, most tellingly, prescribed. These tactics have all the hallmarks of agentic thought processes. More of the currently growing criticism of androgyny as a concept is directed towards the latter than the former approach.

Androgyny is based on the assumption that everyone has some feminine and some masculine characteristics, but that these two sides are more or less developed in different individuals. Singer (1976), whose style is communion-based, describes androgyny as a mode of being characterized by 'the embracing of, and easy flowing between, one's masculine and feminine sides, whether woman or man'. Androgyny is 'creative polarity'. An essential theoretical justification of the concept is that it more appropriately mirrors our essential human nature, than do our current, divided selves. Androgyny, then, is 'based on the archetypal fundaments of the human personality, as shown over and over again in the myths people have told each other to explain themselves and their world' (Singer, 1976).

Colegrave (1979) takes her interpretation of androgyny from mythology and Chinese philosophy. Her narrative stresses the productive interplay of the male and female principles in both individual and cultural development. She defines androgyny as 'the realization of a self which is both differentiated and united'. A precondition of androgyny is that the two principles must be assigned equal worth. The polarities are valued for the unity they imply rather than for the differences they reveal. Writers also emphasize that the two principles must be clear and distinct to unite in this way. They are not advocating blurred sexuality. One criticism of androgyny does, however, interpret this as the message. The concept is seen as denying individuals a base in a particular gender. Bardwick (1979), for example, warns against its use as an asexual interpretation of personhood. She sees gender identity as a critical existential anchor, without which individuals would be unable to achieve a satisfactory sense of self, or identity in relation to others. Communion enhanced by agency, and agency mitigated by communion, offer alternative formulations in the spirit of the androgyny literature, which acknowledge and accept that women and men approach androgyny from different dispositions based in different anatomies and significant, associated psychosocial factors. In this sense androgyny would involve claiming the power of ones own gender *and* drawing on characteristics of the other. The individual would be existentially anchored, but have access to alternative strategies of being. Claremont de Castillejo (1973) advocates a similar interpretation of androgyny. She suggests that society's current disorientation is 'probably the product of each sex being *invaded* by the characteristics belonging to the other rather than being consciously and positively related to the opposite in themselves'.

The notion of androgyny described so far has helped me express the shape and dynamics I see in combinations of communion and agency. Enhancing communion with agency, for example, does not mean blurring the two styles of being (or the female and male principles on which they are based) together, but starting from one to incorporate aspects of the other. This model accords with current developmental personality theories which portray growth as the process of integrating initially separate capabilities. It paints a creative, dynamic and liberating picture of human functioning. Individuals can escape from sex role traps if expected links between gender and individual characteristics become more flexible. But I was curious that many of the attempts to take the theory of androgyny further have disappointed this initial promise. It seems we may not yet be ready to develop it *appropriately*. I find some insight on these disappointments in Colegrave's (1979) use of creation myths to illuminate the female and male principles.

Colegrave depicts early matriarchal consciousness as an integration of humanity with the universe. People were community rather than individually motivated, free of ego-consciousness. 'Order' was observed through 'natural', unquestioned laws. The emergence of patriarchy is associated with a vanquishing of the old, undifferentiated consciousness by bringing order and differentiation into the world. Cultural progress was achieved through the development of understanding, freedom and choice. Colegrave cautions against overidealizing matriarchal order: 'So when we are tempted to look nostalgically backwards to the security and peace of the matriarchal consciousness in reaction to the aggression, inequality and oppression of what followed, it is well to remember the disadvantages as well as the advantages of Her rule'.

Without the slaying of the Great Mother, Colegrave claims

'the further development of understanding would have been paralysed, for in the incestuous embrace of the Great Mother, humanity would have remained an unwitting prisoner of nature's ways, never her conscious collaborator. Before the slaying of the Mother there may have been greater peace, greater equality between the sexes and greater harmony between people and nature, but there was little freedom of choice, little understanding, little control and little individuality. There may have been no oppression of one sex by the other but there was also no real possibility of relationship'.

Only through the overthrow of matriarchal consciousness, then, could the male and female principles emerge as distinct polarities, capable of engaging in dialectical development with each other. Colegrave claims, however, that we do not yet know what the fully developed female principle is. Its growth has been overshadowed and constrained by that of the male principle. We therefore tend to confuse the female principle with the Great Mother which preceded it, but this misinterprets its qualities. This is the stumbling block to developing the concept of androgyny further. Until we have adequate conceptions of female, interpretations of androgyny will tend resolutely towards the agentic.

Daly's forceful attack on androgyny as 'a semantic abomination', 'expressing pseudo-wholeness in its combination of distorted gender descriptions', recognizes the current impasse. Her criticism is directed at the writers who are too-hastily filling the void with man-made stereotypes of masculine and feminine, with all their attendant faults. These are used as substitutes for the more fundamental principles which we are only now in the process of fully exploring.

This premature specification of androgyny is illustrated in the agentic approach to its development. Masculine and feminine are clearly distinguished from each other and attributed characteristics accordingly. The concept becomes an accumulation of parts, and all appreciation of the whole is lost. Masculinity–femininity scales are specially designed (Chapter 2 contained a critique of how this is done), and used to assess individuals. 'Androgyny' means scoring high on both masculine and feminine characteristics. Spender *et al.* (1975) and O'Connor *et al.* (1978) provide typical examples of this approach. They found that individuals who are classified as androgynous are also highest on questionnaire measures of self-esteem. They are followed in this ranking by those with high masculinity–low femininity profiles. These results have been used as a basis for advocating a new ideal person— attracting a further criticism that androgyny is normative. Advocates of androgyny run the risk of simply replacing one coercive norm — that of masculinity — with another.

The main danger in the concept of androgyny, then, is that women may accept premature reconciliations of sex differences and abandon explorations of their own natal identity. If they accept androgyny as an idea and as an ideal, their journey of spinning will have been taken over yet again by male concepts and values. Ironically not only women will benefit from their persistence with the more radical inquiry necessary to illuminate the deep grounding of the female principle and its potential parts in androgyny. Only through such developments can men enhance their appreciation of the male principle in dialectic development with its complement. It is certainly too soon (it may also always be inappropriate) to solidify the concept of androgyny when we have only so far glimpsed the female principle which is jointly central to a full understanding.

My conclusions about androgyny have led me to a clearer expression of my conclusions about the communion–agency duality. To combine productively, the two strategies each need to retain their clear, distinct identities. Compromise debases their qualities. The ways of communion, like those of the female principle, currently require the focus of attention to remedy previous neglect.

VI: CONCLUDING REMARKS

This chapter has chronicled a lengthy phase of my (still continuing) journey towards and within feminism. It has taken my questioning of social values much further and deeper than Chapter 2's superficial prods at socially accepted 'truth'. Through an analysis of language, I have illustrated patriarchy's power to shape and dominate the world in which women live, and so numb women's consciousness. Conceptual models of how and why this unequal pattern of relationships sustains itself need to include adequate explanations of women's values, motivations and roles, as well as those of men. Section III explored possible frameworks. Only through combining an appreciation of communion and agency as alternative modes of human functioning with the popular dominant–muted groups analysis, could I bring women alive in this way. Once this had been achieved, it became important, and possible, to look again at women's characteristics and culture with the intention of re-vision.

In its various phases, this chapter has drawn together some of the ideas I found most exciting and illuminating in my journey towards understanding women and their potential participation in the male world of management jobs. There are too many ideas to develop all of them explicitly in the rest of the book, but most will reappear in some form as I make sense of the experiences of a sample group of women managers.

The dilemma of how to behave when faced with a powerful system of norms and values has underpinned this chapter, as an issue both for women reappraising their role and for me as a writer. There are three broad alternatives. We can conform and adopt the prevailing viewpoint, or we can be counterdependent and reject it as an act of rebellion. The second strategy is as conditioned by the established system as is the first. The third alternative is to develop a new base of values from which to accept or reject norms and values independently, and generate new options. Being constrained by patriarchy means women can either accept or reject its values. Only when they appreciate the workings of social valuing processes and explore their own muted tradition can they become truly free to choose. My own development, from reluctant acceptance of male values to complete counterdependent rejection, and on again to finding my own centre from which to choose, is reflected in this chapter's phases. From these explorations I have spun my own brand of feminism. Part of this process has meant rediscovering and revaluing women's tradition, and mixing more with women at work and home. I have allowed communion in me clearer expression. Before, for example, my attention to the social dimensions of work life was often half apologetic or masked as childish playfulness, concealing my concern that these were less important than getting the job done and might be seen, or even be, all I could contribute. (A double devaluing.) Now I feel I have a more secure sense of when, and when not, to be a nurturer. I can now also better understand and deploy my own agentic self. She is very strong after all these years of development, and needs her head at times. But now the communal me usually goes along on these journeys, to balance, moderate and provide a secure core. She reminds me now to mow the lawn and enjoy the sunshine, rather than start straight away on the next chapter. I have worked enough today.

Chapter 4 returns to my initial division of topics into individual and organizational factors, and considers the latter. I have used my learning from previous chapters to determine my approach, and have taken a selective view of organizational life. Several sections explore organizational structures which intervene between the societal and individual levels of analysis on which I have so far concentrated. These show how social values are translated into effects for individuals.

Chapter 4

What places are there in organizations for women?

INTRODUCTION

In this chapter I return to organization life, but from new directions. With the concepts about valuing, power and stereotypes developed above, I can now look more sensibly at how organizations, as the backcloths to managerial jobs, function. Having explored women's characteristics from various perspectives, I also have a solid basis from which to re-integrate them into an appreciation of organization structures. Women now have positive qualities of their own, and the insidious male as norm will, I hope, lose much of its magnetic power.

I shall take the social valuing processes illustrated in Chapter 3 as the deep structure of organizational life. Starting from an understanding that dominant values type male positive and female negative, I shall spend little time on the surface phenomena which have preoccupied many writers. It comes as no surprise, for example, to find that many industries used to operate a two-tier recruitment policy (or that some still do). Women were taken on to do jobs which involved relatively menial, repetitive tasks, with limited, if any, further development opportunities, and were expected only to stay a short time. Men, with higher basic qualifications, were taken on to provide the company's management layers for the future, and developed accordingly. Banking offers a very clear example of this distinction (Ashridge Management College, 1980). Its operation was usually less apparent. In a report asking 'How much progress in the 1970s?', The National Research Council (1980) of the United States reveals similar sex segregation of entry jobs to scientific and technical professions. It seems only consistent, then, that women at whatever organizational level receive less training than male counterparts, and are overlooked as career material whilst male colleagues are offered promotion. The extent of such discrimination is well documented in the literature on women in management and has been one of its main focii. See, for example, Fogarty *et al.* (1971); Larwood (1977); Brown, (1979); Ashridge Management College (1980); Mirides and Cote (1980).

This chapter looks beyond these symptoms of women's relative position in society, to explore how their structural inferiority follows them into organizations despite official equality. Its concerns are the deep-seated processes by which female devaluing is expressed and perpetuated within organizations. Now is a particularly fertile time to explore these issues, because identifiable discrimination has been proscribed by law. The processes may therefore have to go underground to achieve the same effects. Research provides some evidence that this is happening. McLane (1980), for example, reports that women with Master of Business Administration degrees in the United States earn more than male counterparts initially, as employers pay enthusiastic attention to equal opportunities legislation in their recruitment policies. These women are, however, soon overtaken in salary and career progress terms as forces favouring men reassert themselves.

A company's official structures and procedures represent only one system through which significant issues are handled. Informal systems supplement, and often completely override, them. This chapter sets out to discover how these informal systems operate and what factors influence them. Its four sections represent a selective sequence of significant issues. They are:

I: Organizations as clones of social values

I first need to test whether my basic assumptions of male positive, female negative hold for organizational life now, or whether recent changes have moderated this imbalance. I look at the distribution of women and men in two new industries. Male dominance and female inferiority is still the pattern, despite the opportunities these industries have had to start afresh. My next concern is what organizational forms this takes.

II: Understanding organizations' informal structures and procedures

The second approach views organizations from perspectives other than sex differences in a search for independent explanations of their functioning. These provide valuable understandings of organizational life in terms of power and uncertainty management, but once I map men and women back on to the frameworks achieved I get the same shape — women 'naturally' at the bottom. (Only now I have alternative definitions of 'natural'.)

III: Strategies for gaining organizational acceptance

I next consider the implications of this organizational shape for women managers. Three broad strategies for pursuing job satisfaction and career progress are identified. I find, however, that strategies which help an individual survive usually sustain men's, and male values', pre-eminence.

IV: Dimensions of power

Finally, I draw together the various aspects of power covered in this and the previous chapter to arrive at a compendium of the power dimensions available to women managers.

I: ORGANIZATIONS AS CLONES OF SOCIAL VALUES

Two studies provide material from which I can answer my first question about women's current places in organizations. They investigate whether recently established industries copy the sex-structuring and segregation which are well-documented in more established employment sectors. The computing and travel agency industries as examples do repeat these traditional social values of male positive, female negative.

Dasey (1981) concludes that the computer industry 'despite being only 30 years old, reflects the male/female hierarchy found in the rest of the economy'. She explains this patterning as the result of twin factors: previous education and training, and employers' negative attitudes towards women. More women tend to enter the industry in lower level training positions which offer fewer opportunities for progression. Reviewing relative success rates on Manpower Services Commission's TOPS courses in 1979, for example, Dasey found that 28 per cent of computer operators, 24 per cent of programmers and only 18 per cent of systems analysts were women. The percentage of women drops as one goes up the status and opportunity hierarchy of computing jobs.

Differences in previous education explain some of these recruitment differences. The National Computer Centre selects applicants for training on the basis of aptitude tests rather than academic qualifications. More girls are accepted into their punch and operator courses because as a group they score significantly better than boys at 'clerical ability' and 'accuracy', and many have previous keyboard experience, having learnt typing at school. Boys in the sample of 500 applicants examined scored marginally above girls in the remaining three (of four) parts of the test: general reasoning; spatial ability; and observation and clear thinking.

After initial training, women continue to do less well than male counterparts. On successful completion of the Computer Centre course a higher proportion of men than women enter programming, a route which opens the door for more job opportunities later. Women tend to congregate in data preparation and control which allows little further skills' development or career advancement.

Dasey notes the irony of this sex segregation if we look at the industry historically. The first American computer programmers were 100 young women graduates originally hired by the United States War Department to do manual calculations, who were then assigned, somewhat arbitrarily, to programming one of their early computers. At this time (1944–46), Dasey claims, male engineers and scientists were preoccupied with developing the technological hardware. Only once programming the software emerged as a 'creative' profession did men take it over. This is another example of the process Kanter (1977) calls 'status levelling', through which designations of sex or status shift to bring them back to an equilibrium of male–high, female–low, should this basic pattern be disturbed.

Mennerick's (1975) picture of the United States' travel industry supports Dasey's conclusion that new industries soon become sex structured in very familiar ways. Using a Trade Directory, Mennerick looked at the distribution of women and men in the industry in terms of job — whether management or sales — and agency type and size. Whilst there are roughly equal numbers of male and female employees overall, management tends to be men's work (66 per cent of management staff were male), and less prestigious sales women's work (64 per cent of sales staff were women). Even where women do appear as managers they tend either to occupy organization positions lower than men if in large companies, or to be found in small and less prestigious agencies where there are fewer career opportunities. They also tend to be employed by female top management, except when the latter are in prestigious agencies and are as unlikely as their male colleagues to hire women employees.

Having provided credible evidence that women do tend to fill the industry's lower status positions, the author reaches a curious position by the end of the paper which shows how insidious 'status levelling' as a pattern of thought is. Mennerick is drawn by it to explain any apparently discrepant data in terms of female disadvantage. For example, high concentrations of male employees were found in sales departments in small and medium size travel companies. The author suggests that these roles may, therefore, be more significant than their title implies, providing a wide range of activities which are both intrinsically interesting (such as escorting tours) and better grounding from which to move into management. Because men do them, these jobs must be more important than they sound. Sex again determines judgements of comparative worth.

However circular some of the arguments, these studies illustrate the power and pervasiveness of mechanisms which persistently act to locate women at the bottoms of organizations and in low status jobs. Fundamental social values are cloned anew in fresh organizational forms. The impetus for the most recent wave of the Women's Movement is said to have been women's discovery that even in the newly generated Human Rights movements of the 1960s they were second-class citizens (Coote, 1978).

II: UNDERSTANDING ORGANIZATIONS' INFORMAL STRUCTURES AND PROCEDURES

In an attempt to step outside some of the potential circularity, I shall next join several authors in looking to concepts other than sex, to explain organization functioning. There is now a growing movement away from explaining women's lives in terms of their supposed personal characteristics and towards appreciating how situational factors inhibit their development (for example, Riger and

Galligan, 1980). These alternative perspectives complement previous chapters, as they explore structures, which mediate between the general social patterns and personal needs on which those focused. Attempts at conceptual detachment, however, often misfire because there is such a close correspondence between female and the new factor proposed to explain low organizational status. Nonetheless the concepts developed do offer fresh insights, and demonstrate that the situations they highlight are common but not unique to women.

In this section I shall consider the three topics of informal organization networks, the organization's opportunity structure, and relative numbers of different social types of employee as they illuminate the positions of women in management. These three perspectives overlap, but each has a significant contribution to make to an appreciation of how organizations work.

Informal organization networks

The formal organization of organization chart and official procedures is only one system by which tasks are achieved and people arrange their lives within companies. To understand the organization fully I must also appreciate the part played by the informal organization, a powerful and largely voluntary network of relationships and expectations, which supplements, often undermines, and sometimes completely contradicts, official structure. Failure to achieve membership of this informal system cuts an individual off from significant aspects of organizational life, and can have severe job and personal consequences.

It did not take writers about women managers long to observe that women are often excluded from this informal network (Richbell, 1976). They have dubbed it 'the men's room', and refer to 'the locker room syndrome' to capture assumptions that many of the informal network's transactions are carried out on exclusively male territory, and that there men achieve a closeness which women employees cannot hope to emulate or penetrate.

There are several facets to the informal system as it may affect (excluded) women managers. Informal channels are important for transmitting information and arriving at decisions. Exclusion may mean poorer job performance and a failure to understand how organization norms are translated into practice. This is particularly significant to new organization members, or those entering a new job function. Schein's (1979) work on organization socialization points to the essential part played by social contacts as sources of guidance and information during the first six months of organization membership. Newcomers who are not welcomed into the informal network must learn everything for themselves.

The informal system is one of the organization's main mechanisms for coping with uncertainty. It helps individuals evaluate each other and establish viable relationships through which official work can be achieved. Within the informal network shorthand signs of suitability develop to supplement official, but difficult to assess, criteria of performance and trustworthiness. Whether one plays the approved company sport and engages in the right style and amount of drinking are commonly standards in these terms. Informal behaviour judged against such notions contributes to senior personnel's assessments of subordinates as suitable, or not, for promotion. Where women remain outside informal networks they are less likely to be noticed, even if they perform well in formal terms, and to receive help with their development.

Aspects of relationships which are out of place in detached professional exchanges are expressed and resolved through the informal system. More intimacy or hostility can be displayed unofficially than formal channels allow. Facilitated by private 'languages' of exclusion and inclusion, informal transactions involve important exchanges of influence, reciprocity and conflict. Individuals who cannot draw on this basis for knowing, and being known, may be denied others' cooperation with job tasks, especially at higher organizational levels.

Through these various functions the informal network carries important feedback to individuals about their performance, acceptability and prospects. Being outside its channels of communication cuts the individual off, both from direct feedback, and from understanding the meaning of signs which do penetrate outside the system. A manager may, for example, not appreciate what being invited to a particular meeting denotes, or realize the implications of staying away.

'Outsiders' may also be deprived of feedback on directly work-related issues because they do not share other organization members' base for handling potentially difficult face-to-face situations. Several studies have found that women employees receive less frank performance appraisals from male superiors and colleagues than do male counterparts. The evaluators are unwilling to broach such topics openly for fear of how (female) recipients will react. The women therefore function in a protected environment with restricted opportunities for learning, and their self-evaluations may progressively diverge from others' opinions of them. Material from Waetjen et al.'s (1979) questionnaire study supports this possibility. They matched the self-concept ratings of male and female middle managers against their supervisors' ratings of 'success'. They were surprised to find that 'unsuccessful' female managers reported the highest self-concept scores — followed in sequence by successful men; successful women and unsuccessful men. Whatever the nature of the judgements being made — supervisors may not recognize or rate very highly the strengths the women value, for example — these results show marked mismatches between individuals' self-evaluations and others' evaluations of them which are likely to lead to significant conflict and frustration. Whether women managers are sufficiently self-confident to perform well is a common question in the literature. This material adds a further strand of complexity to the tangled web of others' and self perceptions.

Various factors have been proposed to explain why women, in particular, are excluded from informal organization networks. These systems operate to control uncertainty, and people who are 'different' introduce risk. Women may be excluded purely on these grounds. Richbell (1976) and Kanter (1977) suggest more sex-specific explanations. Many of the informal networks's norms and conventions have been developed in all-male communities and settings. Men may find it embarrassing and potentially restricting to reveal these activities to women's scrutiny. The married male manager who mixes with women *as equals* at work may find this has disturbing implications for his relationship with his wife. The latter is expected to be less tolerant — and 'jealous' or 'suspicious' — of her husband's participation in mixed sex activities than in all-male gatherings. His acceptance of women as peers at work may also bring the balance of their relationship into question — particularly the division of responsibilities at home and his spouse's own interests in paid employment. By restricting women's participation in organizational life, men may therefore be helping contain anxiety about potential conflicts and changes in their own personal relationships.

The element which is missing so far from an appreciation of women's participation in the organization's informal network is their interest in being members. The evidence I do have suggests that women do not value the informal network as highly as do men, or place such a high priority on membership.

Several studies show, first of all, that women do not make such a clear distinction between the two systems — they tend to see the organization as a whole, rather than split into formal and informal parts. As splitting and separation are aspects of the agentic principle, I might expect women to differ from men in this way. A study of Reif *et al.* (1975) provides supporting evidence. They asked 341 participants on management development programmes (only 55 of whom were women) to evaluate certain aspects of their work environment on a range of descriptive scales (such as good . . . bad). The questionnaires covered six concepts, half relating to the formal and half to the informal organization. Yet again research revealed more similarities than differences between men and women. Overall, however, they did perceive the organization differently. Women tended to see it as an interrelated whole; men to distinguish between its formal and informal aspects, and to attribute to the former more potential to impact and reward individuals' behaviour. Looking at more specific responses, women had more positive views of the formal organization and its ability to satisfy their needs than did their male colleagues.

These and similar results suggest that women have to learn a distinction between formal and informal organization, because it is not based on their experiences and needs. Once they have done so, they consistently assign the former more importance than do men. This difference may help to explain why women, in contrast to men, tend to comply more strictly with the official terms and conditions of employment than with its informal demands. Women's lifestyles also affect this pattern of priorities. Women typically want other activities in their lives besides work and take on domestic responsibilities (see Chapters 6–8), and so keep more strictly to official working hours than do male colleagues. They are less likely, from their own volition, to mix socially at lunchtime or stay late in the evening for a drink or game of squash. These are the very times and activities which create and maintain informal network membership. As in Chapter 3, I find that women are partly excluded (from informal networks) by men but also express independent values of their own which have similar effects.

Indications that someone is not accepted into the organization's informal system, and evidence of negative repercussions, are difficult to identify and demonstrate concretely. Richbell (1976) is led by this problem of proof to a near-apology that most of the factors she identifies seem 'trivial' in isolation. She argues, however, that they add up to a significant total picture. Their apparent triviality is part of the very nature of the phenomena she discusses. The maintenance of a relationship *is* the accumulation of minor, often repetitive, exchanges. An analysis of informal organization networks summates these individually inconsequential illustrations to form a total picture in which women's abilities to participate in organization life are undermined by their exclusion from a primary organization communication system.

Position in the organization's opportunity and power structure

My second perspective on organizational functioning overlaps with and complements the first, by illuminating the nature of relations between formal and informal systems. It interprets the organization as a whole as a structure of relative opportunities and power. Before going into detail, I need to introduce the research study by Kanter (1977) from which this (and the third) framework was derived.

Kanter's 'Men and Women of the Corporation' is a particularly valuable analysis in the genre of deliberately non-sexist explanations of organizational functioning. The 'disturbing' introduction of women to new parts of the organization is taken as an opportunity to make the system and its functioning transparent to itself in ways it has never been before. Kanter's concerns are the quality of working life, and organization effectiveness generally. She identifies the modern corporation as essentially dehumanizing. Its boring routines, incipient anxiety, unequal power distribution and lack of any sense of inherent worth, she says, adversely affect people of all types and at all levels. Rejecting the popular argument that having more women in organizations will 'humanize' society, she set out to look for more fundamental understanding of the factors involved in sustaining current patterns:

'This concerned me, because my evolving theory held that as long as organizations remained the same, merely replacing men with women would not alone make a difference. I needed to offer alternatives. . . . I wanted to see the nature of organizations, not men or women as individuals as the villain of the piece, for I was convinced that men were just as much bounded and limited by organisation systems' (Kanter, 1977).

This motivation led Kanter through a series of research and action projects in a large company which she calls Industrial Supply Corporation, or 'Indsco' for short, to write an anthropological 'ethnography' of the corporation. The result is a rich enjoyable exploration. As Kanter's 'women of the corporation' are secretaries and male employees' wives, as well as female executives, she does some unusual justice to women's more traditional roles, and writes these into the functioning of the organization as a whole. In my use of her material I shall concentrate on its implications for women managers, and the conclusions I can draw when I map female–male back into Kanter's initially sex-independent analysis.

Kanter's main argument is that many of the negative stereotypes of women in organizations can be explained in terms of three structural 'facts' of organization life: opportunity, power and relative number. Only 'relative number' is wholly new here. It is developed as this section's next framework. Issues she identifies as 'opportunity' and 'power' have already received some attention, and will therefore be dealt with relatively briefly first.

Opportunity and power are intermediate factors through which potential difficulties in the organization's formal functioning are resolved informally. The distribution of opportunity and power reflects the organization's attempts to handle uncertainty and risk. Kanter uses the two terms as closely similar concepts. I shall deal more fully with the former, broader, analysis here.

'Opportunity' in Kanter's analysis is expressed as potential for movement up the organization hierarchy. With this concept she copies (but with little apparent criticism, to my regret), the concern with career success which pervades many large companies. Some indications of success are inherent in the job itself, which brings with it a particular pathway of opportunity. Women and minority groups often find themselves in jobs with short or no opportunity pathways (although this may result partly *from* being the kind of jobs which these social groups do). More interesting as factors influencing opportunities are those characteristics of individuals by which they are judged suitable, or not, to be employees initially, and later for particular positions in the hierarchy. As this process of evaluation is inherently risky, observable signs that someone is 'the right sort of person' become important. Uncertainty about assessing the vague qualities of 'character' and 'trustworthiness' is controlled, Kanter suggests, by achieving homogeneity, principly in terms of social background and characteristics and of organizational experience. Larwood and Blackmore (1978) similarly conclude from their studies that 'people groom for leadership those with whom they enjoy an in-group relationship'. Again, agentic, control-oriented strategies dominate.

Circumstances of increased threat — when an organization is new, for example, facing a rapidly changing environment or recruiting for highly visible and responsible senior management positions — heighten uncertainty and risk. In such cases the value of common organizational experience is negated, and the importance of personal characteristics dominates. Under threat, organizations recruit from people known to them and emphasize conformity in training, management practices etc. They cannot tolerate the diversity represented by 'unusual' employees. Women executives are manifestly unusual, falling into 'the category of incomprehensible and unpredictable' (Kanter, 1977), and are therefore generally excluded from organizational participation.

With the above concepts, an organization and its tasks can be mapped as areas of relative risk, and predictions made about its staffing. Well-controlled bureaucratic organizations in stable conditions, experiencing low uncertainty, should be able to tolerate more heterogeneity amongst personnel. Ironically the 'order' they have achieved may act to prevent them doing so. 'Deviant' employees are likely to be clustered in low uncertainty areas — more routine functions, for example, or staff and advisory jobs which can be by-passed if necessary. Senior management positions would be high-risk in these terms.

Kanter describes in detail the various responses to low opportunities. She suggests that whilst the attitudes women in low mobility organization positions express are often seen as characteristics of them as individuals or of women as a group, these 'can more profitably be viewed as more universal *human* responses to blocked opportunities.' She finds support in other research results, which disconfirm sex-role stereotypes but can be understood as the consequence of low or high opportunities. One study finds, for example, that work is not a 'central life interest' of (male) factory workers, and another that it is for (female) professional groups.

Responses to low opportunities include disengagement from work; depressed expectations; low commitment; looking to peers or subordinates (rather than superiors) for social recognition; developing an 'anti-success' norm of peer solidarity and conservative resistance. If any low opportunity group member does eventually achieve promotion, they are likely to be rejected by their former peers for breaking anti-success norms. Individuals will tend to cope with this rejection by identifying with higher organizational levels, and are consequently unlikely to represent the needs of members of their previous network once in their new position. Criticism along these lines is frequently directed at women, members of racial minority groups and the formerly poor who move on to senior positions. For the original low opportunity group concerned this could be seen as a self-defeating strategy. Kanter's analysis offers a fresh perspective on the dynamics of this process.

Kanter's use of 'power' rests on an individual's ability to move up the organizational hierarchy. Because of this I think it the weakest element in her model as it is too unawarely dictated by organizational norms as they stand. Kanter defines power as 'the ability to get things done, to mobilize resources, to get and use whatever it is a person needs for the goals he or she is attempting to meet'. Power means commanding scarce resources in the organizational system, and is thus heavily dependent on one's position in the organization's opportunity structure. Managers in Indsco were judged by their ability to provide subordinates with access to opportunities.

Kanter suggests that many of the negative stereotypes of women in organizations can be explained in terms of their powerlessness rather than as direct consequences of being women. She suggests that women often get trapped in cycles of powerlessness, and that the 'mean and bossy woman boss' stereotype is a 'perfect picture of people who are powerless'. In Indsco, twin factors which contributed to women's lack of perceived power were doubts about how far women would progress up the organization, and a widespread belief that if they did progress they could not take anyone else with them (a common practice within the company). The preference subordinates in the company expressed for a male boss is therefore interpreted as 'natural' by Kanter, as it expresses a preference for attachment to power.

These analyses of opportunities and power have appeared circular in places — but so are the processes they are examining. An interpretation of the organization as a structure of opportunities and power, engaging dynamically with its environment, helps us appreciate the processes through which women become trapped in low-status organizational roles.

Relative number

Kanter's analysis of relative number is, for me, her most significant contribution to the women in management literature and I shall therefore describe it at some length. She draws on material from various contexts to develop a multi-dimensional understanding of what it means to be a 'token' woman. Besides its intrinsic values, her framework offers fresh interpretations of the woman manager chestnuts of 'fear of success' and 'the queen bee syndrome', attributing them to situational factors rather than individual character.

Kanter combines several strands of theory and research data to develop a model of how the relative number, or proportion, of socially and culturally different people in a group influences interactions. Her own illustrative material is based on a detailed exploration, through interviews and observation of training sessions and informal gatherings, of the experiences of women sales workers in one Indsco division. Kanter's central thesis is that 'groups with varying proportions of people of different social types differ qualitatively in dynamics and process'. These differences reflect the effects of contact between the categories represented. The main social types to which the theory refers are 'salient external master statuses' such as sex and race.

The model identifies four potential group types according to their relative proportions of constituent social categories. It is assumed that only two categories on a particular master status dimension are present in any one group. *Uniform* groups have only one kind of person in the terms stated. *Skewed* groups, the ones on which I shall focus, have a large preponderance of one type over another, up to a ratio of about 85:15. The few group members are called 'tokens', although there may be more than one according to the size of the group, because they are often treated as representatives or symbols of their category, rather than individuals. The third group type is called '*tilted*', indicating a less extreme distribution of member types, a ratio of perhaps 65:35. The minority in this case are more likely to affect the culture of the group as a whole.

Finally, the *'balanced'* group has a membership around 50:50. Other structural and personal factors rather than the status on which there is balance are likely to influence group functioning.

The skewed group is the type which interests me here, as it is one now being encountered by large numbers of women entering new (for women) functions and levels in organizations. I shall consider, in turn, the meaning of being a 'token' within this model; three perceptual phenomena which follow from their proportional rarity; and the consequences of these phenomena for majority members' behaviour and tokens' reactions. I shall draw my illustrations mainly from the experiences of Indsco's salesforce. Supporting evidence from other research contexts in which a blind person, male nurse or black is the token is also quoted by Kanter.

Not all groups in which one or a few individuals differ from other members fit the theory of skewed group processes. Necessary conditions are that the characteristics on which group members differ should be ascribed and permanent rather than temporary; easily identified; highly salient to individuals; and should carry with them sets of assumptions about culture, status and behaviour. In terms of the official purpose for contact — a management meeting, for example — they may appear to be auxiliary traits, but their importance is heightened if members of the majority group have a history of interacting with the token's category in very different ways from those required in the present situation. The numerically dominant group is assumed also to be culturally dominant over the other. Because they are virtually alone, a token represents their category to the group whether they choose to or not, and can never escape these auxiliary traits. Their role is therefore partly, if not largely, symbolic.

This model of interaction bridges the gap between societal and individual level theories. It shows how sex role stereotypes and social dominance and subdominance are translated into experience for the individual through intermediate structural factors such as group composition and norms. In doing so it helps explain some of the conflicts between conclusions drawn from these different levels of potential analysis.

The interaction dynamics of skewed groups emerge from three phenomena in dominant group members' perceptions of tokens. These perceptions — of visibility, polarization and assimilation — subsequently influence relations within the group, pressures the tokens experience and their responses to these pressures.

Visibility

Because of their rarity, tokens stand out. The Indsco salesworkers reported several repercussions of this visibility. Their performance was closely monitored by colleagues and superiors, so they did not have to compete with male peers to be noticed. But the attention they received was selective. They felt vulnerable to having any mistakes overvalued, and seemed more often to be remembered for secondary factors, such as dress, than for competence or achievements. There were also extra symbolic consequences to the sales-workers' behaviour. They were evaluated as examples of how well women could do in sales, and against standards of how women should behave.

The women's two main responses to these circumstances were over-achieve-ment and attempting to limit visibility by avoiding public events. The first strategy is particularly risky as outstanding performance by a token reflects poorly on dominant members' competence, and can incite retaliation. One common form of retaliation directed towards women executives is labelling them 'hard and aggressive'. 'Fear of success' can, then, be interpreted as a reasonable 'fear of visibility'. This label offers yet another new slant on this now much contested behaviour pattern, but has no startling new implications. I find it interesting, however, as further evidence that fear of anything would be more acceptable to many contemporary American women writers than fear of *success*.

Polarization

To maintain the boundaries between themselves and tokens, dominant mem-bers exaggerate the differences between the two social types. This can make those who are in the majority more aware of their commonalities, as well as serving its initial function of isolating the token. Ironically the token's presence may thus be strengthening rather than moderating the other group's culture. Kanter identifies several ways in which polarization found expression amongst the Indsco salesforce. In the presence of token women, men would often exaggerate displays of aggression and power. They brought sexual innuendo into potentially neutral activities such as training exercises, as if testing out the women members' responses. Women were sometimes cast in the role of interruptor of 'normal' activities, being asked whether it was acceptable to continue in their presence to swear, talk about football, use technical jargon, go drinking, tell jokes etc. In doing so, majority members were drawing, and probably heightening, the cultural boundary between themselves and the token. Often the tokens were thus placed in the position of giving permission for behaviour they could not engage in themselves.

The token position is characterized by several double-binds like this. At one level the individual is offered alternatives, and therefore apparently has free-dom of choice — to allow or not to allow others to swear, for example. But at a more fundamental level of analysis, there is no real choice. Both alternatives amount to the same thing — complying with a role with predetermined charac-teristics, in this case the role of interruptor of normal activities. The individuals are in a double-bind because, whichever option they choose, they are con-strained by a system of rules created and enforced by others. One way to exercise 'real' freedom is to recognize and disengage from these rules by invoking an alternative framework of values and action. Kanter's data give no indication that the women in Indsco tried such strategies. This is not, however, proof that they did not. Disengagement can go unnoticed or be misinterpreted when viewed from within the dominant frame of norms.

As tokens are nonetheless official members of the group, and party to its secrets, majority members have some need to know how trustworthy they are. Within the salesforce this need was expressed in 'loyalty tests' through which tokens could qualify for closer relations with dominants. Tokens were encouraged either to join in negative comments about other members of their category, establishing themselves as exceptions to the general rule, or to allow themselves and their category to be laughed at by the group. Often, if women objected to laughing with the others in this way they were accused of lacking a sense of humour.

As there are too few tokens, by definition, to form a viable counter culture, they can respond to polarization by either accepting isolation and remaining an audience to some of the group's activities, or by trying to become insiders. One of the prices of becoming an insider is that they are then under considerable psychological pressure to adopt group norms towards other women, particularly excluding them from membership. The pattern of behaviour if often called the 'queen bee syndrome' (Staines *et al.* 1974). It is typically interpreted as a sign of women's hostility towards each other, and used to confirm negative aspects of female sex-role stereotypes. An appreciation of the interaction dynamics of a skewed group give it an altogether different flavour. Women who earn tentative acceptance as an insider in a skewed group jeopardize their position and survival if they associate with other tokens.

Assimilation

The third perceptual process that dominant group members engage in is assimilation, a tendency to distort tokens' characteristics to fit generalized stereotypes held of this social group. If several tokens are involved their behaviour may modify the stereotype. Whilst their proportion remains marginal, however, it is easier for the dominant group to retain the generalization and distort perceptions of, and responses to, tokens accordingly.

One form of distorted perception is the 'status levelling' referred to earlier. Status is misperceived up or down to match the sex of the person involved. This brings situational status into line with the individual's 'master' status and social type. Indsco's women sales-workers were often victims of mistaken identity. They were most often mistaken for secretaries, assistants, wives or mistresses of male colleagues. Correcting these initial misperceptions took time and was a socially awkward process. The women were also sometimes aware that, despite such efforts, a legacy of the first impression remained. They might, for example, be expected to do secretary-type tasks even though they had now established themselves overtly as a peer sales-worker.

With its accent again firmly on stability, the dominant group can deal with token members by encouraging them to take on roles compatible with stereo-

typed perceptions of their social type. Kanter identifies four popular generalized roles in which women may become trapped: those of mother, seductress,
pet and iron maiden. She sees the latter as a contemporary category of last
resort. If the token woman will not accept one of the other three she is
identified as tough, dangerous and worthy of suspicion: 'You're not one of
those women's libbers, are you?' was a sure sign that the designation was being
brought into play. Again women are caught in a double-bind. The immediate
options of accepting or rejecting stereotyped roles allow them no freedom or
independence. These four roles are some of the 'places for women within
organizations' this chapter set out to discover. They are particularly influential
as early role options — for a new entrant, for a woman newly entering traditional male areas or in situations of high perceived organizational risk— but lose
some of their imperative over time. The stereotypes may, however, sometimes
operate to exclude women from organizational membership altogether. Heilman and Saruwatari (1979) found that being 'attractive' was a disadvantage for
women applying for traditionally male jobs. Judges gave 'attractive' men and
'unattractive' women the highest ratings on the selection criteria of ambition,
decisiveness and rationality.

The positions of mother, seductress, pet and iron maiden (and other variations on the same themes) effectively encapsulate women, highlighting
apparently auxiliary aspects of their behaviour and precluding them from
displaying task-related competence. Certain roles, particularly that of
'mother', also associate their holders with emotionality, for which they then
risk criticism. The woman manager may consider it easier to accept or go along
with stereotyped roles than continually to battle against them. This could be
interpreted as a dysfunctional use of the communion strategy, which seeks
harmony with its environment rather than trying to control it. One result will be
a certain degree of self-distortion as she accentuates certain aspects of her
personality and suppresses others to match the role's specifications. In this
context, long-term relationships take on a special significance. Through them
the manager can escape stereotypes and becoming more an individual, herself.
The strategies of avoiding contact with strangers and the general public, and of
staying with one company, often attributed to women managers, make such
developments more possible.

The framework of skewed groups illuminates the much bemoaned dearth of
appropriate role models for aspiring women executives. Many available, but
unacceptable, models come from stereotyped roles and women's private
accommodations to structural organizational dynamics. Not only are token
women isolated within the skewed group, but they are under considerable
psychological pressure to identify with the dominant male members rather than
with their own social category in all settings. Any development of a joint
women's consciousness with potential to impact the dominant system will
therefore be inhibited, despite apparently significant numbers of women
spread across the company as a whole. The dynamics of skewed groups help

explain how the dominant culture maintains its own stability, and resists the potential for change women employees are said to represent. Dominant group, male, members work with essentially distorted and limited notions of women which they (men) have largely helped to generate. The three processes Kanter identifies in effect 'dramatize' the dominant group culture in response to its penetration by tokens. Paradoxically the presence of a few tokens does not pave the way for others, and may even have a directly opposite effect.

Bardwick (1979) and Novarra (1980) join Kanter in arguing that simply having more women in management jobs will not solve the many problems of unequal opportunities. Particularly if women are introduced singly, the consequences are likely to be considerable personal stress for them, and very little moderating impact on the organizaton concerned. All three writers argue that a 'quantum' number of women is needed to significantly change organizational cultures. Kanter also insists that they will need to be clustered rather than spread across organizations, effectively changing the dynamics within work units from those of skewed to tilted or balanced groups, and allowing women a significant impact. There may already be industries, occupations or departments in particular companies in which this is happening. What the tipping point beyond which the proportion becomes influential is, and how this significant increase could be achieved practically, are testing questions.

Reviewing the organization's informal structures and procedures.

The aspects of informal organization functioning explored in this section have charted a world which is largely outside the reach of legislation and official company policy. Its mechanisms handicap women managers in their day-to-day experiences of work and their chances of career progression. I have concentrated on the consequences of women's exclusion from informal networks of communication, their relatively low status in the organization's opportunity and power structures and their numerical scarcity.

These aspects of organizational life are primarily concerned with reducing uncertainty, risk and attendant anxiety. Women managers represent threats in these terms against which controlling action is taken. Informal processes become particularly active and influential when official structures cannot handle risk. Decisions about whether to recruit women or to promote them within management are prime examples of organizational crises which call the informal processes discussed above into play. Bartol (1978) identifies these and other critical incidents during career development as an organization's 'potential filtering points' for women. Organizational life can thus be seen as a continuing sequence of acceptability tests. The fundamental issue of acceptability is not resolved, as one might expect, by initial recruitment or rejection, but is invoked afresh each time risk in some form is heightened.

The next section considers some implications for women managers' behaviour of the frameworks developed so far.

III: STRATEGIES FOR GAINING ORGANIZATIONAL ACCEPTANCE

So far in this chapter I have examined some of the informal and structural processes which restrict women to particular positions within organizations, and limit their repertoire of behaviours. I now move on to consider some strategies women can adopt to increase their chances of acceptance, and how other organizational members can help them.

Strategies for survival and progress within organizations

We can identify three broad strategies through which women pursue organizational membership. The three options, which are not mutually exclusive, are: taking advantage of stereotypes of 'feminine' characteristics, adopting male patterns of behaviour, and seeking entry to the organization's informal networks (Bartol, 1978). Adopting a strategy need not be a conscious choice, and will be influenced by *both* environmental factors and individual characteristics and abilities.

One option open to women is to capitalize on stereotypes of 'feminine' characteristics by following career paths which demand and reward these qualities. These are most obviously the 'human relations' jobs in organizations such as personnel, but may also be expert or information processing roles which draw on women's supposed expertise in handling detail. The advantages of these roles are that they allow their holders easy assimilation into the prevailing culture, have a history of accepting women employees, and some are currently expanding in importance and may thus benefit their occupants disproportionally. On the negative side these are usually staff positions peripheral to the organization's main chain of command. They carry low levels of responsibility and status, and are thus potentially frustrating. Such functions are also vulnerable to cutbacks during economic downturns. Many career routes in these organizational areas lead eventually to dead-end jobs because their holders lack the breadth of experience, particularly in production functions, considered necessary for senior and Board positions.

This profile identifies further, more specifically job-related, role traps to add to Kanter's list of female archetypes. Being a caring human relations specialist, an information specialist or a good assistant are three common possibilities which may offer intrinsic satisfactions but are potentially self-limiting. Viewed in the broader context of furthering the acceptance of women as managers, this strategy also has disadvantages. It implicitly supports the stereotype that women are unfit for direct management jobs, and perpetuates 'velvet ghettos' of women's specialisms within organizations.

A second broad strategy is to copy male patterns of behaviour and accept their criteria of performance as ways of fitting into organization cultures. This appears to be a popular initial strategy amongst women who aspire to non-traditional jobs. It was adopted, for example, by all of Hennig and Jardim's

(1978) sample of 'Twenty five women who made it'. Copying male behaviour has some benefits because of its precedents, and women can use these examples (although they have seldom been praised by male colleagues) as models. If rigidly adhered to the strategy can lead to senior management success. Again, though, the stereotype of managers as essentially male is reinforced, and so this approach does little to challenge or modify the prevailing culture. Copying male behaviour requires considerable self-management from the female manager, who must keep any discordant aspects of her personality out of the work environment. She may eventually experience conflicts between her sexual and career identities as a result. Difficulties also arise if others disapprove of such behaviour in a woman.

This, too, is likely to be a self-limiting strategy, as the individual is unable to draw on a whole profile of capabilities if some are defensively muted in order to gain organizational acceptance. The members of Hennig and Jardim's sample who reached senior positions only did so once they had paused in career development to reintegrate initially suppressed (female) aspects of their identity. Those who did not attend to this important task remained at middle management levels. I shall return to this issue in Chapter 6. One possibility, then, considering the viability of different strategies over time, is for managers to earn initial acceptance by adopting male standards of behaviour, and later to find ways of bringing more of their female characteristics into play. How such apparent changes may be viewed, and responded to, by colleagues and employing organizations is open to speculation.

The third alternative is for the woman manager to gain acceptance by seeking entry into informal networks. These will usually be those already existing within the organization, although some women may have the opportunity to establish alternative networks more closely related to their own needs. This approach is more active and self-directed than the two above. It allows the individual more scope to create and present their own images of themselves rather than be defined and trapped into other people's expectations. It may be used as a complement to either of the two previous strategies. In fact, unless a manager does somehow cultivate an informal network, their official authority will be continually threatened by alternative, more insidious, influence processes. As previous sections have demonstrated, entering the informal system in a personally authentic role may still be very difficult for women. If they are unsuccessful the attempt to do so may even damage their position in the company. If they are successful, the benefits are likely to be long-term rather than immediate, and maintaining membership will require considerable time and energy over and above that necessary for immediate task performance.

Other organizational members as helpers: sponsors and mentors

Whichever strategy they adopt, women managers stand to benefit if other organization members help them with the necessary learning about organizational life and about themselves. Studies consistently find that a high proportion of women who are successful in career terms have received such help (Hennig and Jardim, 1977; Roche, 1979). Other organization members take on the roles of mentor — coaching the individual in aspects of the job, organizational life and career development strategies — and sponsor — proposing the individual, often against opposition, for promotion within management. This chapter's earlier discussions showed that these activites are more important to women managers than they are to their male counterparts. Where women are excluded from informal systems based largely on male meanings, mentors can help them learn about this largely 'foreign' culture. By giving personal approval for women's appointment to senior jobs, sponsors reduce the organization's felt risk by taking responsibility on themselves. Mentors and sponsors may also help tokens achieve acceptance into the dominant culture as individuals rather than as stereotypes, by bridging the gap between the two social types. One of the most important functions they can perform is encouraging others to evaluate tokens in terms of competence and achievements rather than auxiliary traits.

The analyses of dominant-muted groups in Cahpter 3, and of skewed groups in this, also show the risks involved in being a mentor or sponsor to women employees. Powerful social norms prohibit dominant group members from developing relationships with those in lower status categories. Doing so is therefore a deviant, and risky, strategy, requiring considerable organizational and personal security.

The processes outlined in this section act to moderate some of the negative effects of informal organizational systems on women managers' lives identified earlier. They have thus served as a reminder that how women perceive and respond to aspects of organizations as they find them will also shape their experiences of work.

IV: DIMENSIONS OF POWER

This chapter would be incomplete without some attempt to draw together the many diverse aspects of one of its and Chapter 3's persistent themes — power. I shall take this opportunity to expand the concept of power beyond its normal range of application. My basis for doing so is a realization that traditional definitions are based on agentic, controlling assumptions about the world and fail to appreciate alternative modes of functioning. Here I shall also explore strategies based on communion to achieve a more comprehensive appreciation of power. This expansion of the concept provides a framework which includes female points of reference, and values women's traditional characteristics. It therefore serves as a map of the varied resources on which women can draw.

The model developed here was motivated by my dissatisfaction with four basic assumptions on which current defintions of power are typically based. These are that power is competitive, a matter of individual ownership, motivated towards control and expressed through doing. These assumptions show the overall bias of notions of power towards agentic strategies and interpretations. In supplementing this view, I shall draw on alternative, communal

108

assumptions that power can be cooperative, based in joint ownership, directed towards influence and expressed in individuals' quality of being.

My thoughts about power currently take the shape of a four-dimensional map which incorporates communal interpretations alongside more traditional, agentic notions. This is portrayed in Figure 4.1. Its four dimensions are: power over others, structural factors which contribute to power, power generated with or through others, and personal power. I have drawn on a varied pool of sources for my individual elements; for the time being they are arranged in no particular order except under their respective headings.

Over others

coercion
reward
ability to access organizational
    rewards (and punishments) for others
formal/positional/legitimate
expert
referent/charismatic

Structural factors

centrality to organizational tasks
handling uncertainty and risk
relative number
visibility
power through difference/new
    perspective

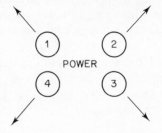

Personal power

competence
wholeness
self-esteem
autonomy
definitional sensitivity and capability
stamina/resilience
change and regeneration

Through/with others

informal networks
politics
coaching/mentor relationships
being attentive to wider community
    issues

Fig. 4.1   Four dimensions of power

The first dimension of the model, power over others, relies heavily on traditional notions of power as a personal resource deployed to control the environment. It takes French and Raven's (1959) classic typology of five bases of power, and adds a further factor. These are generally conceptualized as operating in relationships, are to do with power or authority over or in the eyes of others, and tend to be judged by an ability to influence decision-making in one's own favour. The five initial power bases are: coercion (being able to punish others); reward; expert; legitimate (resulting from ones position in the formal authority hierarchy) and referent (personal appeal, often referred to as 'charisma'). I have added a sixth related factor, which is an individual's ability to access organizational rewards (or punishments) for others. This ability is particularly important in large organizations with centralized reward systems, and was identified in two of the organizational analyses mentioned earlier as a critical precondition of successful leadership (Bayes and Newton, 1978; Kanter, 1977).

If taken as all there is to power, this first dimension of the model encourages spurious notions of personal independence. Power is seen as something that an individual has, ignoring the essentially relational nature of the processes involved. To understand leadership we must also recognize followership; the expert role relies on others seeking and being ready to use information. A woman manager, for example, may be unable to exercise the authority of her official position if subordinates decline to act as followers. Chapter 2's case study of the new team leader, Dr. A (Bayes and Newton, 1978), shows how this happens in practice. Members of her team cast Dr. A in a role of low authority and rejected any initiatives that were incongruent with this definition. The relationships depicted by the first dimension are essentially asymmetrical.

The model's second dimension covers structural factors, aspects of their position in the organization which influence individual performance. Under this heading I incorporate many of this chapter's themes. Individuals may derive power through helping the organization handle uncertainty and risk, performing functions which are central to the organization's critical tasks, or epitomizing the organization's ideal type of employee. Within most organizations there is a need and room for risk-taking as well as for risk-reduction. Being innovative, visible or sufficiently different to represent alternative perspectives within and to the organization can all contribute to an individual's impact on the company and stock of personal credit. On this dimension none of the factors are inherently 'good' or 'bad'. (Such assumptions are made about the first dimension's bases — that being an expert is an advantage, for example — perhaps unjustifiably.) Whether a strategy such as risk-taking is acceptable and appropriate is dependent on the context and on outcomes judged against specified criteria. I would, therefore, expect more variation between individuals and over time in assessments of whether particular circumstances constitute access to structural power.

The third dimension reflects power achieved by and through relationships with others. The relationships in this case are essentially symmetrical, in contrast to those of the first dimension. This aspect of power contravenes traditional deficiency assumptions which interpret power as a finite quantity. Unlike 'zero sum' notions (that if I get more, you must necessarily end up with less), it allows for the possibility of creating power through relationships. Membership of an informal network which exchanges information, makes decisions and creates shared meanings is a major source of this sort of power. Other examples are forming liaisons to influence particular organization issues; coaching relationships in which one individual's growth and development is fostered by another's; and 'social support', empathic relationships which act as buffers against stress. This facet of power particularly contravenes the popular notion that power and achievement are individual possessions. It originates in a wider attunement to the interests of the pair, group or community. In relationships with this communal, synergistic emphasis, it is inappropriate to apportion credit. The power resides in the relationship and is not owned by any individual member. Even if the parties to the relationships feel comfortable with shared ownership, however, others may interpret their collaboration from competitive assumptions. Traditional female roles such as secretary and assistant are typically devalued by attributing the achievements of the partnership to which they belong to their bosses. Because of tensions generated by participants as well as observers, this dimension of potential power is often fragile and unstable. In crisis it is likely to revert to competition.

Fourthly, I use one broad heading to cover various individual aspects of power. This includes skills from which we derive the competence, mastery, self-esteem and autonomy which are our bases for independence of being or doing. The sense of personal self-worth these concepts attempt to capture illustrates a theme of 'wholeness', which I have included separately. This is not used to advocate some ideal type or the resolution of all potential conflicts. Rather it means an individual valuing and having access to all their characteristics through an aware, but not necessarily easily pleased, self-tolerance. My concept of self-esteem is therefore grounded in self-validation rather than the ease in public relationships which confidence denotes.

I have termed a further source of personal power 'definitional sensitivity', to indicate abilities to appreciate how situations and characteristics are being defined and valued by others; be independent of these definitions; resist those which threaten our own integrity; and offer others new definitions as appropriate. Mangham's (1978) dramaturgical model of interactions offers several potential roles in this process. We can, for example, be scriptwriters rewriting the parts in the drama, or producers or dramaturges criticizing the workings of the play. Skills in such roles give us a new power to shape rather than be shaped by our social environment.

Drawing on my earlier profile of women's characteristics, I have also included stamina, resilience, abilities to undergo change and regeneration, and intuition based on understanding others as potential sources of personal power.

In the process of mapping power, I have set previously muted characteristics and activities alongside more traditional, agentic notions. In doing so I suggest that we re-value communion-based resources in our conception of power. The expanded framework presented here incorporates assumptions that balancing the four dimensions' very different qualities affords the individual flexibility, choice and the potential for effectiveness. Sex role stereotypes tend to restrict women to a very narrow range of strategies — mainly power through or with others, and the expert role of power over others. To move beyond this trap they need simultaneously to expand their capacities to operate agentically, and to retain, but re-value and possibly redeploy, their traditional power bases.

In Chapter 7 we shall see how a sample of women perceive themselves in relation to different power bases, and the dilemmas which many are now facing about how to operate effectively in management jobs.

Summary

This chapter set out to discover what places there are in organizations for women. I found that they tend to be in low responsibility, low-risk jobs, outside organizations' informal communication systems or trapped within stereotyped roles (some based on female archetypes such as seductress, pet, mother and iron maiden, others on task-related images of expertness in human relations or handling detail), and explored the processes which keep them there. Informal networks foster conformity to reduce uncertainty and risk, and so exclude unusual categories of employee such as women. The dynamics of already skewed social groups act similarly to keep women in the minority. Whatever an organization's official policy, these processes operate as the deep structure of their functioning, and are resistant to change because they are self-perpetuating. In the chapter's last two sections I explored some of the implications for women managers of operating within this context. Adding communal notions of power to our well-established agentic definitions particularly expands their options for achieving both an acceptable sense of self and successful job performance.

The next four chapters bring the issues discussed so far in this book to life in a profile of thirty women managers. Chapter 5 sets the scene by describing the research study in which they participated.

Chapter 5

Towards a profile of thirty women managers

From previous chapter's explorations I now appreciate some of the factors in their complexity which might affect women in management jobs. This chapter pauses to review how, given the themes already introduced, I can approach the stories Chapters 6–8 have to tell about the sample of thirty women managers I contacted. Section I suggests various ways in which the research data are relevant to the book's evolving interests. Section II provides a more practical introduction to the research study itself. Sections III and IV gives some background information on the two industries studied.

I: REFLECTION

In previous chapters I have argued strongly that there are differences between fundamental female and male paradigms of being and doing. The world of employment, particularly at senior organization levels, is dominated by male values, needs and perspectives on the world. The research material which follows explores the impact working within this system has on women's lives. The chapters deal in turn with three major areas of choice which face women managers: their relationship to work; issues of personal identity and lifestyle options. The sample's choices were not made about one of these three dimensions at a time, but about their lives as wholes. The chapters therefore interlock as different ways into a complex, multidimensional appreciation of their experiences. Through their exploration we turn with more focus to the core questions which motivated this book:

What do you want?
Who are you?

The next three chapters map and consider the possibilities, and show how some of them are translated into the lives of a small sample of women. These

112

findings do not indicate an ideal way of being a manager or becoming success-ful; rather they provide signposts for women who want to be awarely in charge of their own choosing (see above). Several significant themes run through the research material. Four particularly carry forward the book's main developing issues about women and work covered more theoretically in earlier chapters.

Choice is one continuing issue. This is expressed not only in terms of the range of available options, but in indications of what choosing means to women managers. Most of those in the sample draw to varying degrees on several different options to make up a whole life. They are not motivated by greed and wanting to 'have it all' as some current writers suggest (Pogrebin et al., 1978). In fact the managers would reject these writers' basic assumptions that women have no rights to engage in activities outside the home and so must justify doing so, that the major options of work and family are mutually exclusive, and that any choice must be pursued single-mindedly to prove acceptable. In their various alternative approaches to each of these issues, women are rewriting what it means to be a manager, a marriage partner and/or a parent. These are steps towards developing new meanings of personhood. By exploring women's perspectives, on this and other topics, the research data contribute to the woman-centred processes of spinning and re-vision mentioned in Chapter 3.

The need to reflect on appropriate meaning for this group is highly apparent in relation to the data's second recurring theme, those aspects of the world of work which managers find appealing and therefore seek. Employment itself, job challenge, achievement and promotion are the common elements here. It does, however, become difficult within the language and concepts currently available to make the distinctions the women managers favoured between how they see men pursuing these as goals, and how they themselves view them. In their attempts to articulate their perspectives, and my attempts to portray these adequately, we again see spinning in action. Applying it to the notion of 'career', which in many ways represents the essence of male values as they are enshrined in employment, is particularly illuminating of femaleness.

The third theme is the other face of the second, the difficulties women experience because they work in systems dominated by male values. Resulting manifestations of tension range from expectations that their future career progress will be impeded by discrimination, to conflicts of personal identity and lifestyle arising from their daily encounters with settings which deny some of their basic values. I also report how they sought to handle these various pressures.

Throughout the accounts of managers' experiences and views are expres-sions of an alternative base of values which underpin the choices they make and their acceptance, rejection or re-vision of current norms of employment. These are further contributions to a female paradigm, a base for women-centred notions of being. Their focus here is on employment, but as I have already noted they are about whole people rather than parts. Whilst the roots of these values can be seen in the profile of women's characteristics and culture intro-duced at the close of Chapter 3, they are at the same time newly created through personal confrontation with a world based in largely conflicting assumptions. Engaging in this challenging process has led many managers to a sureness and sense of accomplishment which is all the more precious because they can evaluate it against their own hard-won criteria of worth.

In these four themes I am again struck by the duality which pervades women's experience; the duality between their reactions to a public world shaped largely by men and their abilities to draw on, and act from, alternative, independent perspectives of the world. Listening to women's authentic voices, I cannot accept that they have mislaid their truths, as some writers claim. Rather they are finding new combinations of assertiveness and appropriate muteness through which to express them. One of my main tasks at the end of Chapter 8 will be to draw together the elements of a female paradigm expressed in these managers' accounts.

II: INTRODUCING THE RESEARCH STUDY

It is now time to introduce the research I did myself. Doing so involves stepping back both conceptually and personally. The ideas and frameworks developed above are my current thoughts. When I actually embarked on interviewing my sample of women managers I was still in what I have called my reform feminist phase of exploration. As I have indicated, this was not the most engrossing part of my journey, it was therefore a welcome change to embark on some action. I translated my reaction against the reformist literature I had read into a strong resistance to letting the assumptions it expressed bias my expectations and subsequently the accounts interviewees would give. I therefore adopted a relatively uncommitted approach to research, especially trying to avoid any assumptions that women are discriminated against.

To set the scene for my research material, I shall now briefly describe how I found my interviewees — by no means an easy task — and approached them from a research point of view.

Getting the project started: plans and practice

The research study was funded by the Equal Opportunities Commission who wanted to derive a profile of women who succeed in management jobs. There were limits of money and therefore time. My dissatisfaction with the available literature's biases and the fact that much of it had come from the United States made me see my work as an exploratory research study in a largely un-developed field of interest. I chose the route of qualitative research, seeking an in-depth understanding of the relevant factors for a small sample. Thirty people could be contacted comfortably within these specifications. I then had to decide who these people would be.

Women in middle to Board level organizational positions were approached as they would have a longer-term perspective on working than would junior colleagues. The sample in fact spanned roughly two working generations, and the material was additionally illuminating for this reason. I decided to interview women only, but to split the sample between two industries to provide a basis for comparison, and thus an enriched context within which to interpret findings. Several factors played a part in the selection of book publishing and retailing as target industries. There was little available material on women managers in these sectors. A large percentage of the workforce in each is female, yet women have only recently penetrated significantly into senior jobs. Their numbers there now, whilst not high, are sufficient to provide a suitable sample. I did not want to interview lone women in male preserves as I wanted my sample to represent a more 'ordinary' constituency of people.

Having identified a broad target group, I made further decisions to limit some of the potential influencing factors to a manageable range. This was particularly important in view of the small sample size. I decided:

(1) To interview only in and around London. This would reduce the significance of whether or not managers were geographically mobile to a minimum and allow other, less obvious, factors to emerge. This was easily achieved in publishing; but it proved unwise in retailing as it excluded an important section of women in general management positions and/or working in stores. Four of the eventual interviewees in retailing therefore lived outside London.
(2) To impose an upper age limit of about 45 years. This would reduce potential differences between generations which are likely to be a source of great variance.
(3) Wherever possible, to contact several people in any one company as a guide to the relative importance of individual and company influences.
(4) Deliberately to restrict the number of personnel people in the sample to avoid weighting the study towards this traditional stronghold of female employment.

Finding the sample was by no means easy, and was an informative process in itself. Although I was often told by informants that there were 'a lot of women about' in their company or industry, it was difficult to track down suitable individuals. I suspect that the high visibility of the few women managers there were created this effect. Having discovered a contact in a particular company (sometimes a woman manager, sometimes a helpful member of the personnel department) I would ask her or him to suggest (other) potential interviewees. I then tried to recruit each participant over the telephone in order to explain fully, and answer questions about, the origins and aims of the study. This was not always possible and a few interviewees were first contacted by post. One person politely declined to take part because she did not want to be 'spot-lighted' in this way. I was however, generally welcomed with interest and openness. Interviews were recorded and full transcripts produced as a basis for data analysis.

My aim in designing a set of basic questions to ask in interviews was to understand something of people's lives as wholes, with the focus on work both for career and everyday aspects. I did not assume that they saw being a woman as important, and so left the one direct question about this until *they* had been able to tell *me* what was significant. Above all I wanted to allow the managers to express their own accounts in their own terms. I used my questions as a base from which to develop joint conversations. My prompt list was as follows:

Please describe your career history since starting work, telling me the important factors as you go along.

What factors have hindered your progress?

What factors have helped?

How do you see your work life developing?

What are the pressures on you in your current job?

What are its satisfactions?

What effect has being a woman had, if any? (Modify if they have already mentioned.)

Please tell me about your life outside work.

What are your main pressures and satisfactions outside work?

What advice would you give to a young woman starting out and wanting a satisfying working life?

Analysing the data

The interview transcripts were analysed through a process of content analysis for their themes at various levels. As I read through, sorted and categorized the managers' accounts, profiles of significant events, influencing factors, attitudes, commonalities and differences gradually emerged for particular individuals, groups within industries, each of the two industries and the sample as a whole. My approach to the material can best be described as 'immersion' (Marshall, 1981), and involves trying to appreciate inherent patterns rather than impose preconceived ideas on the data. The researcher becomes an interpreter rather than a manipulator, concerned with capturing other people's meanings rather than testing hypotheses.

One emphasis in my use of the material is the identification of common experiences, attitudes and feelings. I have used these as a basis for developing theories and models. Within the sample there was, however, considerable diversity, and a second main concern has been to reflect this adequately. Doing so is particularly important in an exploratory study such as this. It not only gives the full flavour of the data, but suggests issues which may be revealed as more significant even dominant, for a larger sample or one in a different situation. At both a sample-wide and an individual level, the account contains many apparent contradictions. That between not being ambitious and reaching the Board of ones company, for example, or wanting to lead 'a balanced life' but becoming highly involved in work. These conflicts are real and alive, and I have portrayed them fully rather than try to explain them away. They are part of today's upheaval in the norms about a woman's place, and are being expressed as dilemmas in many individual women's lives.

Who the sample were

Employment Half the women managers worked in book publishing and half in retailing. They came from five publishing houses and four retailing companies. All but four women worked in or around London.

Age The two industry subsamples were similarly distributed in terms of age — between 26 and 50 — which was roughly assessed from career histories unless volunteered. This increases the comparability of material between publishing and retailing.

Life stage Four women were married with children; a further eleven were living with husbands or long-term partners; twelve were single and living alone, and three single and living with parents. Three women were divorced or separated. Only one attributed this to her involvement in work. The others said it would have happened anyway — 'it's the kind of person you are'.

Job functions Of the fifteen women in publishing, eight were in editing or publishing, four in selling rights (explained below) and three were in more general management jobs.

Six retailing interviewees were in staff management, five in buying and four in general management posts.

Organizational level Management grades cannot be easily defined or assessed and only rough evaluations were made. In publishing, four interviewees were on the Boards of their companies (Directors), a further four at one level below this ('senior' management) and the remaining seven in 'middle' management positions.

Two women in retailing were on Boards. A further eight were classed as 'senior' management, using a wider definition of one or two reporting levels from the Board to reflect longer management chains in this industry, and five as 'middle' management.

There is a tendency in woman manager research to redefine the designation 'manager' in order to find sufficient female respondents to fit the research criteria. This process of adjustment has some validity, but can be taken too far. In one study I heard of, for example, the women 'managers' were all Girl Guide leaders (at a special convention) and were considered comparable to a sample of men in industrial companies. In my sampling the only qualification to the meaning of manager is typical of the industry rather than a reflection of double standards. Some of the editors in publishing had very few subordinates for their organizational level, but the same is true for male colleagues. The different distributions of men and women in industry raises testing methodological issues for those intent on comparison-based research. I favour doing research where women are, rather than creating artificial parity.

Industry profiles

I shall next briefly introduce the two industries with my early impressions of them to give a flavour of the settings in which managers operated. Further material will be covered in later chapters as it relates to specific topics under discussion. Particularly interesting for me was seeing industry themes reflected in individuals' personal styles and lifestyles. It seems that people both sought out an industry and company which suited them and were then further shaped and developed in its image.

III: BOOK PUBLISHING

Small and friendly

The book publishing industry is relatively small, employing approximately 16,000 people, most of them in three dozen or so larger publishing houses. It is also highly centralized geographically, most of the activity being in and around London. I was told repeatedly how friendly, even when it comes to competition, the industry is. There is a lot of communication between companies— 'It's an incestuous business, everyone knows everyone else anyway. It's like a big club' — and a 'grapevine' operates making it reasonably easy for personnel to move around with the industry. Either an employee is enticed away to another company at the latter's initiative, or word gets around that someone is dissatisfied where they are and an alternative offer is made. One manager cited as evidence that women were becoming fully accepted in the industry, that some had recently been 'head-hunted' by other companies.

Opportunities in the industry

The main career path in publishing is the editorial 'route'. Other possible jobs are sales, administration, the selling of rights (to publish a book abroad or adapt it to a different medium such as a film), promotions, and work in production or service departments (photographic studios etc.). These are all discussed in more detail in the next chapter.

Publishing is a popular occupation relative to the number of jobs available and does not have to attract applicants. The industry is difficult to enter and has the reputation of being highly selective, employing a relatively high percentage of Oxford and Cambridge graduates and people with first-class degrees. Several interviewees mentioned the pressure towards having not only a degree, but the right kind of degree. Some of the four who had not been to a university mentioned this as a disadvantage. Because of this exclusivity, traditional advice to women has been to go into publishing as secretaries and 'work their way up'.

Several of the sample had entered in this way. Some would still advise others to do so, while others are very wary about this option.

Recent expansion

The publishing industry has expanded a great deal over the last 10–15 years, with some areas, for example Rights, growing even more rapidly than others. Both these general and particular trends have been especially important in opening up opportunities for women. Most people interviewed say that growth has now slowed, and career progression will take longer for everyone in the future.

The industry's attitudes to women

I was told that originally even secretaries in publishing were male; but there have long been places for women in the industry, even if in small numbers and relatively poorly paid, at middle and senior levels. Traditionally, low pay at all levels has contributed to creating pockets of jobs which are unattractive to men, particularly detailed editorial jobs. Also, it is accepted that women are as likely as men to have the desirable characteristics of creativity and skill in dealing with people which many jobs demand. This openness to women has been greater in some areas than in others. For example, the selling of rights (which originally was often the part-time responsibility of a director's secretary) and editorial jobs have largely been done by women.

Many of the sample feel that today publishing is one of the industries most free from prejudice against women — 'it's just not a terribly sex conscious business'. Others were more qualified and said that some prejudice remains at unconscious levels and that some companies and individual male managers are more 'enlightened' than others. This group, though, see a further 'opening up' now in progress, particularly of senior jobs and places on Boards. Many of the sample are pacesetters in moving into these positions.

IV: RETAILING

Competition, urgency, mobility

In Great Britain approximately two million people are employed in retailing; it is therefore a much larger industry than publishing. The close family atmosphere of the latter is missing, and there also appears to be relatively little communication between companies. The interviewees contacted for this study came from large companies involved in general retailing.

Interviewees' accounts of life in the industry were underpinned by a notion of 'the rhythm of retailing'. They talked in terms of the demands of business, competitiveness, urgency and the constant need to make decisions and to keep up to date. They gave an impression of demanding, but usually exciting and enjoyable, work carried out at speed. Capable individuals can as a result be swept along by this tempo and develop their skills more rapidly than they might in more placid surroundings. One manager identified 'the urgency of retailing which means you have to learn as you go along and gain a lot of experience quickly', as a significant contributor to her own progression. With this emphasis on current performance, university degrees are not essential entrance qualifications, and may even be frowned on as being too impractical. Only six of the managers in retailing had one.

The structure of the industry — relatively independent stores distributed throughout the country and reporting to regional management and then central head offices — has two particularly significant implications for employees at management levels. Firstly it offers large numbers of relatively independent jobs, opportunities to run one's own show and prove one's capabilities. It also, typically, means considerable geographic mobility, especially in the early years of training.

Opportunities in the industry

Retailing tends to attract few graduates and much training is 'on the job', supplemented by short courses. Most large companies have some form of management trainee scheme which offers basic training, plenty of shop floor experience and an opportunity to decide where to specialize later. Usually there are two 'streams' to this initial training, leading to later development in either commercial or staff management. These are closely interrelated, but roughly handle the business and people aspects of retailing life respectively. They are described in more detail in the next chapter.

Places for women

Commercial management has traditionally been a male, and staff management a female, specialization. The top in-store jobs for each stream are 'store manager' and (his subordinate) 'staff manager' respectively — recognized as the 'father' and 'mother' of the unit. Commercial progression routes are longer and lead to a much wider variety of senior management positions than do those via staff management. Whilst equal opportunities policies have removed these sexual demarcations officially, they have left significant legacies. Many managers argued that someone has to be able to handle the personal problems of staff, who were assumed to be mostly female. Approved characteristics of an employee in commercial functions still closely match sex role stereotypes of masculinity, confronting women who enter these areas with conflicts about personal identity.

The sample feel that because there have always been some places for women in retailing, life is easier for them now. Currently more women are going into general management and commercial posts, and getting through to senior levels. This, in its turn, is increasing acceptance and further opportunities for women. A few interviewees said that retailing is particularly good territory for capable and ambitious women, since similar men are attracted to more glamorous jobs, such as marketing, thus reducing the competition. Several women commented on the impact of equal opportunities legislation on their companies. Initial over-reaction in trying to promote token women too quickly was said to have given way to promotion based on merit independent of sex. Many, however, were more qualified in their assessment of progress, and reported the high visibility of women who fail. A few, especially those in commercial jobs, agreed with one who said: 'It's still a man's world. There are opportunities, but they aren't equal'.

Chapter 6

Women managers' employment histories, and futures

In the next three chapters I shall draw on my conversations with thirty women in retailing and book publishing companies to reflect on their experiences of working as managers. I shall focus in turn on the core dimensions which emerged from our discussions:

(1) their evolving relationship to work, including expectations they held of the future;
(2) issues to do with personal identity; and
(3) their overall lifestyles.

This chapter deals with the first of these topics.

I must emphasize again that managers seldom separated these three dimensions of their lives as neatly as I shall do for the purposes of writing. They insisted that their lives were, or should be, wholes rather than separate parts. Choices in any one area were made in the context of considerations in the other two. These chapters are therefore essentially complementary. They move round three perspectives on the managers' lives, illuminating them from these different vantage points. This means that there are some necessary overlap and points of contact between them.

EMPLOYMENT HISTORIES

In writing about women managers' employment histories, I am faced with a dilemma. It is easy to slip into the language of careers, which values promotion and progression, and expects people to plan ahead to achieve opportunities and status. But these assumptions were held by very few of the managers I interviewed. All but one, it is true, have worked continuously, but most have definitely not been looking upward as they did so. Only a few have deliberately progressed by building up a career in the way we traditionally use that word as a meaningful sequence of jobs each preparing their holder for greater responsibility in the future. Instead their typical orientation has been each job as they did it. In this chapter I shall therefore use words such as 'career' and 'ambition' with caution, and shall sometimes pause to spin new meanings for them, which are more compatible with female experiences.

This chapter deals with factors which have shaped the employment histories of women currently in management jobs. A sequence of five phases is identifiable in the managers' accounts of their working lives. The phases can be characterized in terms of the managers' attitudes to work, their companies' opinions of them as employees, and key factors which both parties consider in their decision-making. Table 6.1 summarizes these factors within the phases' framework. Interviewees also saw personal characteristics as highly significant. As these are not specific to particular phases, they are gathered together in a separate section towards the end of this description.

PHASE I: EXPLORING THE OPTIONS

My first area of interest is how the managers got started. Whilst their experiences are highly varied, most had an early phase of experimentation. As the differences between retailing and publishing are relatively marked, I shall consider the industries separately to begin with.

Few of the publishing sample showed a clear initial sense of direction in their early employment choices. Twelve of the fifteen did other things before entering the industry; these ranged from 'odd jobs' for several months, to another occupation altogether for eight years. In all, five had always wanted to go into publishing or to 'work with books'. As direct entry is difficult to achieve, they deliberately acquired secretarial or other related experience until they were able to get a 'proper' publishing job. The rest either did not know what they wanted to do, or tried other options first. These experiences helped them shape ideas of what they were looking for. Interesting work, promotion prospects and acceptable ethics (some felt uncomfortable with those they found in the business world) emerged as priorities. Their main motivations were for challenge and satisfaction in the particular job they were doing. If these waned they looked for promotion opportunities, more to provide greater stimulation than for their increased status.

Getting the first job in publishing was either a case of persistence — for the five who had specific intentions — or luck. Many of the others answered job advertisements when feeling unsettled. Ironically the 'persistent' group often accepted relatively menial first jobs as a way in, whilst the others went straight into positions more appropriate to their qualifications.

In contrast to the relatively quiet persistence or search of the publishing employees, most of those in retailing showed a more attacking approach to work from the very beginning. Nine, for example, actively did not want to go to university: they were keen instead to gain experience, 'get working and earning', to prove themselves and not to waste time. There was, too, more initial direction to their employment choices — all but four going either into

124

Table 6.1   Phases in managers' working lives

I: *Exploring the options*

Experimentation
Trying what turns up
Persistence to enter chosen area
Moving to achieve better fit: of training opportunities, job interest, challenges, ethics
Key people and their advice
Developing confidence

II: *Achieving early career moves*

Increasing responsibility and challenge
Promotion, despite company reluctance, via chance factors, persistence, jobs that grew
    under them
Comparing self to male contemporaries
Attracting attention
Mentors and sponsors
Valuable learning experiences: success in challenging jobs, trade union activities

III: *Middle management levels and steadier movement*

Have won acceptance from company
Seen as relatively low risk
Managers guard their acceptance by staying with one company
Managers now limit their own career progression to lead a balanced life, stay with
    meaningful work, avoid job pressure, stay put geographically

IV: *The senior management barrier*

Lack breadth of experience for some senior positions
Women become high risk again
Visibility
Mentors and sponsors
Managers' own ambivalence about top jobs

V: *Senior and Board level positions*

Being exceptional
Satisfaction
Running out of headroom
Seeking more independence

retailing or lines of professional development which were to be their continuing
interest. The main reasons for their choices were being attracted by opportunities for women (in retailing's staff management stream) or by the excitement, dynamism and relevance of work in business, and people they knew or
met who encouraged them towards retailing. Frequent movement is a striking
aspect of these people's early accounts: geographic (including some who had
worked abroad), and between industries, companies and types of work. Moves
had largely been motivated by needs for growth and achievement. Members of
this sample showed an initially surprising extent of taking responsibility for

their own job satisfaction and futures. The most frequent explanations given were moving to take on vague but challenging opportunities or to widen their horizons, and away from companies in which promised training proved disappointing or career prospects seemed limited.

For a significant minority of younger managers in retailing, advancement emerges as a significant motivator in its own right. They clearly did want careers in the traditional sense. The following comments are typical examples:

'I just felt there was no real career structure for me in the kind of work I was doing. . . . I wanted growth in the job, progression, a reasonable income level and a certain amount of independence. The interest was not enough.'

'It became obvious I was going to have to wait for the next job until I was "suitably aged". It would only have been three years, but it seemed a long time then.'

'I soon realized that we were waiting for deadmen's shoes and so obviously I started looking round, because I wasn't particularly wanting to stay in the same sort of position until I was 40.'

This group do indeed stand out from the rest of the sample for their more calculating attitudes to career development. Deliberately planning future flexibility — at work or in general lifestyle — is another aspect of this. Several of this group are now in senior management positions. They particularly identify having stayed single and mobile as critical factors making their rapid progress possible.

What is consistent for the sample as a whole about these early years is the phase of experimentation they represent. Whatever people's initial intentions, they used this as a time to try out options, reject those which seemed inappropriate and take frequently unusual and unlooked for opportunities which then proved valuable. Often apparently trivial experiences, such as chance meetings or conversations, would have radical results in the new directions they initiated. Eventually individuals arrived at a combination of type of work and organizational setting which suited them. Most then stayed with that one company until I met them. During this period many younger people (particularly, again, in retailing) had also learned that they had some competence, and had established an initial commitment to work. One young manager's recollections particularly typify this combination. Surprised at being voted into charge of a work's kitchen (during university vacation work), she realized she could look for a career in management:

'It was the first time I had anything to do with controlling people and organizing things. I must admit I was surprised I was chosen and that was the first thing that made me realize I could perhaps have a career somewhere in industry or business.'

For (now) older managers, positive directions such as these seemed to have come later, if at all. In this early stage they had been concerned about job interest and doing a particular job well. Work in a wider sense then became increasingly important as they were eased up the organization by chance factors, encouraged and developed by the faith of a superior, and either did not marry or did not have children. Theirs was a much gentler, slower establishment of commitment to work. These seem to be typical findings for older women managers and help explain their general lack of official qualifications, and higher age per organizational level compared with male colleagues.

As this phase closes, then, the individual has usually settled on a type of work and in a company which she feels suitable. Some have developed an initial confidence in their competence. Most rate job challenge and satisfaction as prime concerns. A few, amongst the more recent entrants, are explicitly motivated to do well in career terms. In the next phase individuals move on along particular routes of development.

PHASE II: ACHIEVING EARLY JOB MOVES

Before considering how early moves within a chosen or accepted line of development were achieved, I shall describe briefly the various routes people had followed in job function terms. The people I interviewed were chosen to represent the range of management jobs currently typical of women employees. In each industry there were three broad options. In publishing the three streams of development are: the editorial route, 'Rights' and general management. In retailing they are staff management, store management and buying. Some of these career routes — chiefly 'Rights' and staff management — are traditionally women's work. In these streams, at least at junior and middle organizational levels, women are at an advantage over male employees.

Career routes in publishing

Assistant editor → editor → publisher

In general, 'publishers' go out commissioning books, seeking and negotiating with authors, whilst 'editors' take care of the book 'in house', from reception of the final draft through to production. (This distinction between the two jobs is not universal, but will be used for simplicity here.) Movement in this stream is the main form of employee development in publishing. Progress upwards is towards more freedom to commission and create books rather than handle those commissioned by other people. Movement sideways to different subject areas ('lists') is also common. Whilst women have for some time been accepted as editors — and praised for their attention to detail and administrative abilities — their movement into publisher jobs is a more recent phenomenon.

One threat to development for women in the editorial stream is the traditional resistance to sending them abroad, a necessary experience in some areas. They are told that foreigners will not do business with them, that they are more likely to behave in ways which tarnish the company reputation, and that they simply are not tough enough. Often cases of disaster, usually many years in the past, could be cited to substantiate these claims. Attitudes are now relaxing, and some of the potential advantages of women as diplomats are being recognized. It is, however, an area in which women feel watched and on trial, some going to great lengths if given opportunities to travel to prove their abilities and moral fibre.

Secretary → Rights

'Rights' are entitlements to produce overseas versions of a book, publish extracts or adapt it for television or film. Selling Rights is now a significant source of revenue for many publishing companies. Rights business has grown dramatically in importance over the last ten years, and with it has grown the job scope and status of those involved. In most companies, what was the part-time responsibility of a director's secretary grew first into a full-time job, then into a department, and now often warrants a place on the Board. As Rights is traditionally a female job — both because of its historical roots and because (I was told) men, who dominate other areas, accept and like to be sold to by women — many women have been swept up their organizations by these developments. Most of those in the sample who followed the industry's traditional advice to women and started as secretaries, had benefited in this way. A high proportion of publishing's women directors have emerged via this route. Whilst on the whole satisfied with the jobs they now do, one or two of those I interviewed expressed regrets and some anger about the image Rights is acquiring as a bastion of strong-minded females — 'The Rights *Ladies*' as the media calls them. They are already feeling trapped by this stereotype-in-the-making.

General management

A few women in the sample had moved into general management, usually more by chance then by design. Running departments concerned with sales, artwork or book production are typical examples of their jobs.

Career routes in retailing

Employees entering retailing generally start in one of two broad streams of career development: staff management or commercial management. The latter splits further into two branches: sales and buying. Women have long been accepted in staff management and certain areas of buying, such as women's fashion. Their movement into commercial functions is more recent. As commercial jobs offer a wider perspective on the business than does staff management, they are the traditional routes to most (that is all but personnel and some buying) senior and Board positions.

Staff management

Staff management is the personnel facet of retailing administration. As the industry is a service industry, it is, however, much more of a line management function than this term usually implies, and is directly commercial in many respects. The managers said of their role:

> 'You're responsible to the manager of the store for recruitment, training, all amenities, the counselling side and also the deployment of people and of their career patterns within the store.'

> 'The role is very much involved with day-to-day running and the commercial side of what goes on in the store — ensuring that all standards are maintained and that the store is run efficiently.'

> 'When my job was assessed it was recognized as being very much an executive function. The job is very commercial, you can't afford to be theoretical. You have to be very conscious of time, you get caught up in the rhythm of retailing.'

Progress in staff management is from assistant to deputy to 'staff manager' of a store. Promotions usually involve moving between stores of different sizes to take on greater responsibility. Short-term secondments to related Head Office areas, such as training, may also be included. The staff manager reports to the store manager (see below) within the management team, which varies in size from just these two in a small store upwards. The route to staff manager is shorter than the commercial route. The staff manager is traditionally seen as the 'mother' of a store, and the store manager as its 'father'. Regional and Head Office appointments may follow in-store jobs. Some geographic mobility is usually essential to career development.

Store management

The commercial management stream covers more directly business-related activities. One path of development is from sales assistant to running first one department of the shop and then departments through assistant and deputy to store manager, the top in-store job. This route parallels that of the staff manager described above. It leads to a higher in-store position, and brings more opportunities for later promotions to and at Head Office.

Buying

Buying is an alternative path of commercial development. This is a separate function, usually based at Head Office. There are two complementary aspects to buying: one is concerned with design and getting the product right, the other with financial aspects and stock control. Within a buying team for a particular area of merchandise both aspects will be covered. In large companies, individuals usually specialize in one or the other. Progress is initially between different areas of merchandise, both to gain experience and to take on more responsibility. More senior positions involve overseeing increasingly larger product ranges, ultimately having little direct involvement with merchandise.

The sample's early job moves

Having once settled on a company and/or functional area, most interviewees report a phase during which they won acceptance from their current employer. This happened through a variety of means which this section illustrates. This phase is important for the opportunities individivuals have to develop their abilities and to demonstrate their competence to other significant people in the organization. Equally important is the interest the company shows in them by offering job moves, promotions and other learning opportunities. How persistent employees are in their commitment to work and further development also plays a part. Finally there are roles for other members of the organization to help as mentors and sponsors. Again, experiences differed significantly between the two main industry groups. A major factor contributing to this is a difference between their career development practices. In retailing, accepted procedures for training, staff appraisal and career development operate almost automatically. In publishing, procedures are generally less formal and reliable.

By far the majority of moves within retailing were achieved through established channels. Interviewees were moved between stores, taking on greater responsibility as they went. Open discussion of career opportunities and evaluations of performance were typical. Sometimes the opportunities offered had an especially exciting edge to them, such as trying to improve performance in an unprofitable store, or setting up a new outlet. In a few cases development was accelerated by additional training, as companies responded to equal opportunities legislation and applied positive discrimination.

In publishing, in contrast, the majority of job moves at these organizational levels seem aptly described by the much overworked simplification 'being in the right place at the right time'. *Sudden vacancies* and (what I have termed) *'job expansions'* are the two main means by which this happened. Both provide exciting examples of fortuitous employee development. Often vacancies would occur unexpectedly — because someone left or fell ill, perhaps. Sometimes the nominee was a fairly natural choice for the job; at others pure availability seemed to be the deciding factor. Even though they were still at lower organizational levels, these promotions usually represented major jumps forward in career terms, and those concerned initially found their capabilities stretched. One interviewee had only recently been made assistant to a particular management position when further promotion came: 'I went home on Friday night doing one little bit and I came back on Monday to be told, "if you can do that one you can take the whole lot over".' These early experiences had far-reaching repercussions by rapidly developing individuals' skills, giving them a new sense of their own capabilities and improving their standing within the company. In this way they are highly reminiscent of findings from the longitudinal Bell telephone study of management development, that early job challenges were one of the factors most predictive of later career success and of involvement in work (Bray *et al*. 1974).

Being in a job that grew under them (job expansion) was another unplanned, but highly significant, contributor to individuals' development. Several interviewees who had been their company's specialist in an area, found themselves swept along when this increased in importance. Selling Rights has already been mentioned as an outstanding example of a function which has grown publishing-wide over the last ten years or so, particularly to the benefit of women. The growing importance of personnel jobs because of increasing legislation in this area and rising labour costs provides a parallel in the retail industry. There can, however, be some disadvantages to development of this type, mainly because it happens gradually. As the job becomes more managerial its holder may regret losing its intrinsic satisfactions:

'I don't really think of myself as management. It's getting further and further away from what you are making. I feel like a paper shover and I don't actually like it. I cling as hard as I can to doing real things.'

In a few cases where this growth had happened largely unrecognized, the manager's job title had taken some time to catch up with her change in status. This had caused difficulties in relationships with peers and seniors. Lack of concern about appearances or status on the manager's part, or not wanting to make a fuss — 'I wasn't terribly worried what I was called' — contributed to delay in addressing such anomalies.

In the examples of early career moves given so far, the initiative has been mainly in the hands of the company. At least a third of the sample had also actively supplemented these methods by *persistently pushing to be given more responsibility*. This tactic was reported more often by younger than by older managers, and was slightly more common amongst those in retailing than in publishing. It had two main motivations — parity with male colleagues and the need for sufficient job challenge. Looking back many managers felt they might have moved on more quickly at these lower organizational levels. This was not the case for those in retailing's staff management stream who were aided by standard staff development procedures operating in a, until recently, female-only career path. Interviewees in other areas felt companies are more active in promoting male employees automatically. These attitudes represent a barrier against which women generally have to push to be considered suitable promotion material. By taking their own initiatives women can go some way towards countering any disadvantages.

Employees in retailing had the advantage of being able to refer to standardized career procedures to judge whether they were moving more slowly than their male contemporaries, and therefore had a sounder base for taking remedial action. Whilst someone in publishing is now left with vague feelings that she was at a disadvantage, her contemporary in retailing had more evidence at the time against which to check her suspicions, and from which to contest unequal treatment.

> Publishing: 'I still feel that the very fact that I was a woman, management just weren't taking me quite as seriously. That they thought with women you can just leave them there and they will just sit quietly and do their editing'.

> Retailing: 'I got on my high horse about that, not because I wanted to go on the course, but because I objected to the fact that he had been chosen and he was definitely not as intelligent or as qualified as I was'.

Not all those who pushed their companies for promotion drew their dissatisfaction from comparisons with others. Many grew restless as soon as their capacities were not stretched by a particular job, and looked for promotion to offer them new, personal challenges. These statements are typical both of their initial concerns and the tactics they adopted:

> 'I was always knocking on the door saying "what about me?". Your potential doesn't get considered if you don't.' . . . 'I would say "I don't want to do that any more, I'm not being stretched".' . . . 'I used to get bored with what I had and so I kept asking for more responsbility. I always had to go to them.' . . . 'I heard of the training scheme and went along and said, "that's the kind of thing I'm interested in".' . . . 'The company didn't promote me so I went out and got a job elsewhere to force their hand.'

In these examples, enhanced status is scarcely important as a motivation; job interest and making full use of personal capabilities take precedence. These people consistently felt that their abilities were either unrecognized or ignored. All those who pushed their orgaizations to notice them identified 'persistence' as a quality they valued and advised others to develop. Some half-joked that they had been given the moves they asked for merely to keep them quiet.

Other less official or obvious ways in which interviewees developed their job skills during this phase influenced their later progress. Some managers identified key *learning experiences* as important. These make up a not atypical list of challenges through which young workers can develop skills and confidence, and earn the respect of other organizational members. Setting things up, working in a small unit and therefore being involved in all aspects of its operation, and succeeding in a particularly difficult task, are some more commonly mentioned. One distinctive example is the learning that one-third (at least) of the women in publishing have gained from participating in the setting up of local trade union branches within the industry. Some had been willing and others reluctant recruits to office. The benefits were deeper insight into the company and its workings; opportunities to develop negotiating and politicking skills; and having their capabilities recognized by senior managers to whom they would not usually be exposed. Some had not initially enjoyed the visibility the trade union role involved: 'I was suddenly picked out of the heap and dealing with the Managing Director'. A couple felt at the double disadvantage of being seen both as 'activist' and 'emotional woman'. Overall, though, unionization seems to have been achieved relatively amicably in publishing and to have offered the women concerned a relatively friendly, low-threat and not directly job-related arena in which to practice what later turned out to be valuable skills.

Other people who helped foster and accelerate managers' development figure significantly in interviewees' accounts of this stage and are also relevant to later phases. Half the sample made special mention of more senior company personnel who, as *mentors* and *sponsors*, had shaped and facilitated their career progress. These individuals fulfilled a variety of roles — must noteably coaching the manager in job performance, advising on career decisions and nominating her (often in the face of opposition) for promotion. Having their mentors' expectations to live up to had encouraged many women managers to perform well. One manager illustrated how an advocate can spread a high opinion to others in the company. Her sponsor typically introduced her with 'don't be fooled by the exterior, that's a man's brain in there'. She is uncertain whether she thinks this an insult, but recognizes the powerful impact it has had on people's attitudes towards her.

Mentor and sponsor relationships develop most often between boss and subordinate. Most, but not all, of those mentioned by the sample had been with men. Once firmly established, the relationship will often continue as the mentor and their protégé move on to new jobs, and possibly to new companies.

Despite their values, mentor and sponsor relationships also bring risks to both parties. There are the obvious dangers of associating with someone who then falls foul of powerful people in the company. When and how to end the relationship are also difficult issues. Most of the managers I interviewed felt that they had at some time to assert their independence, and combat others' expectations that they had merely been assistants, if they were to carry on successfully in their own right.

Further elements in industry profiles

I can now add another strand to the industry profiles started above. This one captures some of the less formal opportunities available for individuals. Looking back over the career sections, I can see that industry differences are particularly striking. Despite its haphazard formal development procedures, publishing seems to offer more varied informal learning opportunities and more 'help' in this sense to individuals in their development than does retailing. These tend, however, to make ability visible only at a local level. A reputation is established amongst immediate colleagues and unofficially, and will only be more generally known if it becomes a topic for discussion on the company's or industry's informal communication system. Being responsible for publishing 'a good book' or clinching a lucrative Rights deal tend to attract this wider publicity.

Retailing provides rather different learning opportunities, with high official visibility for success. Two contributory elements in this stand out. One is what people referred to as the 'nature of the business' — the urgency of retailing which means that employees with any degree of responsibility have to learn quickly. Many individuals' personal styles matched and mirrored the competitive pace they ascribed to the industry generally. Secondly, responsibility is located in a multitude of small units — departments or stores. As areas of responsibility can be relatively clearly identified and sales provide a speedy measure of performance, 'success' is both clear and visible. Employee evaluation becomes a relatively objective process. It is also easier to identify individual contributions than in publishing, where publisher, editor, author, production and marketing (at least) all play parts in a book's success or failure.

Phase II is a busy one for the manager who will later reach senior levels. She will draw on a wide range of experiences to develop her abilities and confidence, and may find herself moving faster at times than she feels capable of. This growth will, however, stand for little unless other associated factors influence her organization to take her seriously as potential development material.

PHASE III: MIDDLE MANAGEMENT LEVELS AND STEADIER MOVEMENT

There seemed general agreement amongst the sample that once they had passed a certain level and become accepted, life became easier and promotions were more regularly offered to them. For a while, that is. Many of the people I contacted were at this stage or had recently passed it.

Themes of maintaining stability and minimizing risk-taking predominate for both employee and organization during Phase III. From the company's viewpoint the individual has now demonstrated her consistent motivation, both towards work and away from competing alternatives of family and child-care. She has also established her capabilities, and developed credibility on this basis. As the manager comes to represent fewer major uncertainties than previously, the company risks relatively little by offering regular career development opportunities. Managers find themselves more fully integrated into the career system, and feel equally treated as far as promotion opportunities are concerned.

For individuals reaching these levels one of the main issues seems to be not losing what they have gained. It is at this stage, for example, that the benefits of staying with one company become most significant. Having achieved acceptance where they were, whether by quiet or more demanding persistence, most managers seemed reluctant to start afresh elsewhere. There were several indications in earlier chapters that a woman's credibility, identity and confidence are typically more situational and relational than those of male colleagues. They will therefore be less easily transferred to new situations, new companies. Women who step out of or above traditional female positions risk being stereotyped with devalued images (see Chapter 4). Many of the managers in the sample had escaped from these in their own organizations, and established distinctive identities. The benefits of staying with one company were therefore high. It is consistently reported in the literature that women managers favour this tactic for gaining and maintaining acceptance.

One repercussion of staying with one company is that, once the manager is known and accepted, she can bring previously muted aspects of her personality into play. These may be her female-based skills which would not have recommended her as an employee earlier on. Managers may also be motivated by wanting to avoid learning about a new organization. They already lead overloaded lives (see Chapter 8), and do what they can to spare themselves further pressure.

Several members of the sample were highly aware of the implications of their recent lack of movement between companies. They thought it limited their future opportunities, and realized that they could be seen as lacking initiative. One said she felt open to accusations of cowardice. The managers were usually ready to explain and defend their apparent inertia. They did so mainly in terms of liking the people they worked with (which also appears to stand for having developed viable working relationships with them), satisfaction with the company and the opportunities it offered, and job interest.

A second major theme in manager's accounts of this stage was the constant attention they paid to the shape of their life style overall. They placed a high priority on 'leading a balanced life' and made choices which they thought would help them do so. The managers were now more likely to limit their own progress than to have it limited by others. Most of the numerous examples they gave of turning down jobs came during this phase. They were also less likely to push for promotion than previously. Married and single managers explained the issues involved differently. Whilst both groups wanted to lead balanced lives, the latter's notions of what this might mean were often vague and long-term. Married managers typically gave examples of translating their concern into practice.

The managers offered several reasons for the shift in their orientation to work. Some wanted to take responsibility for the speed of their own progression, refusing to be hurried upwards unless they felt fully capable of taking the next move:

'It seems to me that it's a good idea to make things solid. I was scared that I might become fashionable too quickly. I began to do one or two things that were quite good and I was at my most alarmed about my career then. I wanted to prove that I could go on doing something solid and serious and something that continued.'

Many of the group did not want promotion, since the increases of administrative work, job stress, isolation and political mode of operating they saw at the next level did not appeal to them. It was not the responsibility directly involved in the job itself which deterred them, but the demands promotion would place on them to pay more attention to appearances at work, and the repercussions it might have for their lives overall and their views of who they were. This distaste was not, however, sample-wide. A few women had developed an interest in management both as an activity in its own right, and in the opportunities it offered to influence others, which overrode their other concerns: 'The more I see of management, the more interested I become' . . . 'I like being significant . . . what is really rewarding is being able to influence.' One expressed particularly clearly the delicate conflict she faced about whether or not to seek or accept promotion. Her comments may be a more appropriately ambivalent expression of others' concerns too. She admitted that, despite her reservations, she would probably accept if asked to take the next step up: 'I don't want to be a director, that's only a jumped-up administrator. Quite nice, though. My weakness would be if somebody were to say "we want you to be a director", or something like that. I don't think I would have the strength of character to say "no, I really don't want to do that". I think I would want to be able to do it'.

Other considerations managers took account of in their decisions about work were aspects of the balance, and potential conflict, between work and home lives. Rationing out limited energy, and the importance of where they lived, were the most frequently mentioned factors. Many mangers, single and married, did not want to 'risk' (their word) devoting too much energy to work at the expense of other areas of their lives. Most based this concern on a previous phase when life had become unbalanced in this way. Other examples will appear in later chapters, but here is one typical expression of the philosophy and its roots:

'Three and a half years ago I was ambitious, but now I don't want what is left of my social life wrecked. I like to leave here in the evening and forget it. When you get too high up the tree you can't. Your work runs your life and I like a little bit of leisure.'

Some married managers said they could not move geographically because of their partners' jobs, but by no means all were limited in this way. Some women thought their companies made this assumption on their behalf, so the issue was never put to the test. Further evidence that there is no simple link between marriage and mobility comes from the attitudes of many single managers who were unwilling or completely unprepared to move. This was particularly so if they were older and had spent sufficient time in one place to build up friendships and a sense of belonging there. One single manager, for example, for whom promotion would necessitate relocation, said she would require 'lavish assurances' about her long-term prospects before leaving the roots and quality of life she had established. Her comments are typical of people weighing choices during this phase for her clarity about, and willingness to take, responsibility for the outcomes:

'I realize that I would be limiting my ultimate career progress, but then that would be a conscious decision on my part; and if that then means that I continue at this level, that is something I would recognize in making that decision.'

As I interpret the managers' reservations about further promotion I must also consider the values I use in doing so. Some commentators complain that women are over-cautious in the choices they make, somehow personally deficient in terms of drive, ambition, confidence, assertiveness or whatever. They suggest that women be helped to develop these missing qualities. Hennig and Jardim (1978) are particularly blatant in taking this approach. They see all the ways in which their sample of women differed from male colleagues, such as not having future career plans and thinking the official organization system more influential than informal networks, as faults. As usual I am suspicious of the simplicity of such conclusions, especially as they judge women by men's standards. If I listen fully to women's own testimony I form very different impressions. Some of the managers' reasoning which prompted them to pause in career terms may indeed reflect underestimation of their own capabilities or overestimation of constraints on their participation in employment. But far more prominent in their statements were their commitment to perspectives and values which relegate employment to the status of one life area amongst others. On this basis, they are making positive decisions about what they do and do not want to do and be. Many are saying, in particular, that they know what promotion, moving and becoming more committed to work will mean for other aspects of their total lifestyle, and that they do not want those consequences.

It seems, then, that many women at least pause in their careers at this stage. In doing so, some may be addressing issues about personal identity as well as the aspects of choice considered above. Hennig and Jardim (1978) found that, without exception, those managers who went on to senior positions had a moratorium on career development at some time during this middle management phase. They attended instead to aspects of their identity so far neglected — particularly more 'feminine' interests. Some married at this point. This redirection of attention seemed to compensate for previous imbalances in development. Being female was too significant an element of identity to be neglected in the long term, as many organizational roles still encourage women to do. Hennig and Jardim's research data suggest that reintegrating an appreciation of femaleness was not only valuable for itself, but was actually a prerequisite for managers to move on successfully to more senior organizational positions. Taking time out in this way was noticeably absent in the life histories of a comparative sample of managers who did not progress beyond middle management levels. These findings closely parallel the views of

EXPECTATIONS FOR THE FUTURE

I was initially surprised at the variety in people's responses when asked about their likely futures at work. Most of their considerations have, however, already been introduced, and you may know what to expect.

Many managers did not, in fact, look far ahead, nor did they think they should be doing so. Their time horizons might stretch to the next job; as far as they were concerned the future was a relatively immediate phenomenon. As the managers in the sample were than at career phases III, IV and V, limitations on opportunities for women at senior levels were one major concern. Managers in general management and commercial jobs had reported significantly more past and present disadvantages of being women than had those in staff management, Rights and editing. This follows the traditional demarcation between functions relatively closed and relatively open to women. I was therefore surprised that career blocks are seen by as many managers in traditionally female areas as in other types of work. This successful sample's prospects were very mixed and uncertain.

The managers' images of the future fall into three broad groupings, based largely on their satisfaction with the present. Some were satisfied and expected to be so for the foreseeable future; some were dissatisfied or uncertain and wondering how to proceed; and others had decided that they needed to take new directions. Table 6.2 shows the numbers in each category. In the descriptions below, married managers tend towards current satisfaction and wanting to deal with changes if and when they occur, and single managers to more active anticipation of, and inclination to shape, their futures.

Table 6.2   Managers' expectations of the future

Satisfied for the foreseeable future:	
Career minded matched by opportunities	5
Happy in current job or at current organizational level	11
	16
Dissatisfied or uncertain:	
Need more challenge, but current company unable to supply	7
Want to have children and continue working	2
	9
Ready to take new directions:	
Want less involvement in work	4
Want new career path	1
	5

Satisfied for the foreseeable future

Just over half the sample, sixteen managers, were satisfied with the future as they saw it. Five would like further opportunities for promotion and were relatively sure that these would be offered by their current employers. The remaining eleven (some of whom were already directors) said they were happy to stay in their current jobs, or one at the same organizatinal level. A few entertained the possibility of moving up later, depending on circumstances. A variety of factors contributed to their perspective.

In job terms, the managers were satisfied as long as their key requirements of continual challenge, interest and growth were met. Many felt their current jobs provided this anyway. They did not have to look to promotion to renew their interest. Some people in this category said they had difficulty visualizing themselves in the future and had never looked ahead in this way. Others did not like the possible futures they could imagine. The next job up might involve additional pressure or require a geographic move. They might have difficulty staying up to date and competitive as they grew older, and anticipated greater job stress as a result. The managers did not expect employers to tolerate less from them than the high standards of performance on which they had built their reputations. Another contributory factor to staying put was the belief several held that working for too long in senior jobs makes one 'hard'. They did not want to risk fulfilling this stereotype. Concerns about leading a balanced life were also frequently expressed.

Most of the group who were satisfied with their current organizational level were married. Their disinterest in looking far ahead extended to all life areas. They deployed their energies to managing the present, and expected to cope with changes as and when they occurred. It seemed that they did not initiate major changes themselves, although they might adjust elements within their current profile of demands to achieve a more balanced life.

The reasons managers gave for current and expected future satisfaction were partly under their own control and so revealed their values and priorities. These are, however, expressed in relation to an established, often constraining system on which the women's self-determination has limited impact. Some of their proposed tactics are therefore pure rejection of that system's demands — especially of ways they might have to distort their personal identity to survive in higher organizational positions. Interviewees also felt that opportunities for women at senior levels are limited. Their policy of waiting to see what the future brings may, therefore, also be in part a reactive moderation of their expectations in the face of anticipated obstacles. In the past this group, particularly, had concentrated on the present and found that the future had taken care of itself. They were adopting similar tactics again.

Dissatisfied or uncertain

Nine of the sample, just under a third, were unsettled in relation to work. For seven this meant wanting more challenge, but doubting whether their current companies were either able or willing to meet their needs. A few, but not all, had recently reached the senior management barrier. None was yet aware of viable opportunities outside her current company, although all planned to search more thoroughly. Having so far stayed with one company (and, for those in publishing, being reluctant to leave the industry) set limits to possible options. Two managers in this category were married and wanted to combine any move with achieving more independence and flexibility to accommodate home demands. Most of the remainder were single and had less pressure of this kind.

The other two unsettled interviewees were facing rather different choices. They wanted to have children and to continue working. They faced four, at least, basic uncertainties: those surrounding conception, whether they could become pregnant; how their bosses would react; trying to anticipate their own feelings, particularly whether they would want to return to work after maternity leave; and the practical problems of arranging childcare. In these circumstances, issues which come under the general umbrella of 'leading a balanced life' cannot be handled privately as the sample would generally prefer. Because they impact, and therefore need sanction from, the employing organization they must be negotiated publicly. Legal maternity rights are merely one element in a complex jigsaw of relevant factors. Unless one's boss and colleagues are at least tolerant, being a working parent is seen as a daunting task.

Ready to take new directions

Five members of the sample, all but one of them single, were 'looking for alternative ways to spend their time', but had so far lacked sufficient impetus to make a break with what they were doing. Only for one would this mean maintaining her currently high commitment to work. She wanted to try a different professional field but did not relish the thought of losing organizational status, even temporarily. The others were more concerned about reducing the emphasis on work in their lives and achieving a better balance. They echoed many single women's concerns about leading 'half lives'. This group felt increasingly trapped by the choices (towards work and away from other options) they had already made, but had been unable to remedy this imbalance. Most were therefore continuing their current commitment but were alert to opportunities which would allow them to devote less time to work. Sideways or downwards moves and self-employment were the main options contemplated. They recognized, however, that at least part of them was 'hooked' by their current way of life and wondered whether they could make the changes they fantasized.

Several managers in this and the previous category said they 'needed a push' to help them move. When I wrote to the sample a year later it seemed that for some their conversation with me had served this function. Several had moved on to new companies or professions, and I heard rumours of other changes. For a group who had previously been relatively immobile, a lot seemed to have happened in one year.

TOWARDS A FEMALE PARADIGM OF EMPLOYMENT

From this chapter's material I required one contribution to the book's development above all others — that is an understanding of what employment means to women managers. Drawing together the strands of evidence from their accounts, I found I was working towards a female paradigm which has its own inherent pattern. Part of this pattern is its relationship to organizations and their male shapes and norms. But this is only one dimension. I can now start from women's experience to build an alternative, independent framework. Many of the concepts commonly used about men's employment are not transferable. The notions of 'career' and 'ambition' particularly need to be revised radically if I want to apply them meaningfully to women. This closing section moves towards expressing a female paradigm, with the focus on employment. Its main focii are their lives, their approaches to working, and the framework of values within which their choosing makes sense. At the close of Chapter 8 I shall integrate this beginning with what I learn about personal identity and lifestyle.

The managers looked on employment as an arena for self-fulfillment and development. Their main motivations were doing work they enjoyed and which challenged their capabilities. In this sense, their focus was on current job satisfaction, but it was dynamic and open rather than static and inward-looking. An expectation of change was seldom far away.

They expected, and wanted, to be changed themselves by what they did and how well they did it; and they expected circumstances to change around them and that they would adapt to their new characteristics. Most had an orientation which was change-accepting, but only occasionally did they move towards change-seeking. When they did, they usually acted from attunement, intervening to shape situations to suit their purposes. Sometimes they reached out to exert more directive control in the face of counteracting forces. In these activities, too, the core motivation of self-fulfillment was dominant. They wanted to assure themselves opportunities for further challenge and development.

The managers' job interest, being personally (rather than externally) motivated, and their present-orientation combined to shape their work styles. They tended towards involvement and usually performed well as a result. They used good performance as a personal criterion of achievement. One repercussion of this style that most managers had not initially expected was that other organization members recognized their capabilities and rewarded them with the traditional prize of promotion. About a third of the sample, older managers generally, could explain most of their movement up the organization in these terms. They described their progress as 'accidental'. Continuous employment was typically not their original intention, but after that ('just') happened a combination of hard work and being recognized resulted in offers of promotion. They saw and accepted these as new challenges rather than as increased status. Several expressed pride in achieving so much without deliberately trying or having had to push. As one explained: 'Whatever job I have done I have enjoyed and tackled; and somebody has always suggested I move on to something, and really there is nothing more to it than that. That may sound unenterprizing but it happens to be true. It's always flattering I suppose to be offered something.'

Some of the younger generation of managers, having learnt the connection between promotion and increased opportunities, have actively, but selectively, pursued this route for themselves. Those in this group, also about one-third of the sample, have continually paid attention to developing their careers. Again their main motivations are the use and development of their abilities, and to be fully stretched by work. But they differ from those whose progress has been 'accidental' in their sense of themselves in comparison with others. They believe that they are equally capable as, if not more capable than, their colleagues, and have therefore wanted at least to match others' progress. Despite this, most will reject the label 'ambitious' as it is traditionally applied. The job rather than organizational status is still their focus. Their explanations illustrate this priority and show that, once initiated, the process of moving up the organization hierarchy acquires its own momentum:

'I always wanted to get on. I always wanted to be the best in what I did. I could see that I could do things; and when I didn't think I could, I thought "well, I'll have a go at them." And when you find you can, it gives you confidence to go on to the next thing.'

'I am not terribly ambitious; does that sound odd? It isn't because I set out wanting a career, because I didn't. It's more that I want interest and more to do in a day than there is time for. It seems people welcome that attitude in retailing, and you progress fast.'

Ambition is a difficult word. If you mean wanting to do something as well as one can, then women are probably more conscientious than men. If you mean wanting power for its own sake, that's not a quality I think I or my friends in publishing have.'

To complete the picture of how interviewees viewed progression I must remember the few managers in retailing who do accept the label 'ambitious'. The excitement and challenge of achieving evermore responsible management positions has taken on an incentive value of its own. They apply the competitive strategies by which they achieve day-to-day commercial performance to their own career planning. Even they, however, reject jobs that do not offer intrinsic satisfaction. Several also express doubts about the pressures of the lifestyle they now lead, and wonder 'whether the sacrifices are worth it'.

Most women managers, then, are motivated by the job itself and the interest and challenges it offers. Only when these become unsatisfactory do they act on the environment to create new opportunities. This balance of strategies accords with my earlier discussion of women's modes of dealing with the world. Their dominant approach is based in communion, the union-seeking life strategy I described in Chapter 3. It involves acceptance of what is. From this focus on current being, new directions emerge. The managers also have access to more agentic, control-seeking strategies. When the situation fails to match their motivations, they recognize this through dissatisfaction and act on the environment to achieve a better fit. They supplement communion with agency in their approach to planning, too. Some anticipate possibilities, such as conflicts between work and childcare, and think through strategies for coping with them. Much of the planning they engaged in or advocated was not goal-directed, but orientation-shaping in this way. Whatever their balance of agency with communion, most interviewees showed a communion-based awareness of the environment in which they operated, of their 'ecology'. Translations of this awareness into their lives ranged from tolerance of contextual forces and constraints, to battling against them. The group of younger managers who had learnt to value promotion for its own sake represent an extreme in these terms. Agentic strategies appeared to have taken over, and this was usually associated with them losing touch with other values in their lives.

One manifestation of women's environmental-connectedness was the importance the managers placed on relationships. In this chapter I have outlined the many practical benefits of building up and staying with a network of people who accept an identity which accords with the female manager's own sense of self. Valuing relationships is also an expression of communion, a preference for a world shaped in cooperation rather than competition. Women managers accepted their connectedness with others, particularly outside work with husbands, partners and children, and its repercussions for their life-style. To interpret these repercussions only as disadvantages or advantages would be to take too individualistic a view, and would not do justice to their approach. The verity of being linked to others was one of most managers' basic assumptions, and was beyond evaluation, it just was. Several of those at very senior levels regretted that such a sense of interrelationship and mutuality was largely missing from their work lives. They found their jobs lonely and isolated. Relationships were also seen by the sample as a potential source of stability and support, providing a sound base from which to reach out into the future and the changes it would surely bring.

Finally, in these steps towards a female paradigm of employment, I can identify some of the values which underpin women's perspectives. Some can be portrayed as balances between potentially antagonistic tendencies. I have stressed how central the quest for self-fulfillment was to them. Valuing relationships, connectedness and cooperation is its complement. Openness also emerged as a guiding principle. This was best illustrated in their attitudes to the future. The managers believed in taking things as they come, and most only planned to the extent of preserving flexibility to cope with developments. Many prized their adaptability and said that most possibilities can prove enjoyable and challenging if approached with faith. Balance itself was also a value. I shall explore its various meanings further in Chapter 8. So far it has mainly been used to weigh the place of employment in the manager's life as a whole. Balance means leading a multi-dimensional life in which work is only one element.

In these steps towards a female paradigm of employment, looking from the perspective of employment itself provides only a partial view. The managers are continually balancing issues of personal identity and life-style alongside those of work. It is now time to move on to the first of these complementary perspectives on their lives.

Chapter 7

Issues of Personal Identity

Concerns about personal identity, and about clashes between their images of self with other people's images of them, permeated the managers' accounts of their lives. They could refer to few accepted norms about who to be, having departed from most traditional female stereotypes. Being outside normative guidelines gave them many freedoms, but also made it necessary for them to choose, often to create new options, and to handle others' conflicting expectations of who they should be. These processes involved risk, and with it uncertainty and anxiety.

To focus issues of identity here I have selected five interrelated areas in which they became particularly central. I shall move from a relatively general to a narrower focus, along individuals' paths of choice about who they are. The sections reflect choices-in-the-making, but this is not to imply that managers were necessarily conscious or deliberate in their choosing; more often they lived out particular options and then became aware of their implications. This chapter explores:

I: Gender: is being a woman important?

I asked how aware the managers were of gender. Their ambivalent replies show some of the ways they coped with being women in a male world.

II: Everyday problems of being women, and strategies for managing them

This section takes a more detailed and practical look at problems the managers experienced and how they dealt with them.

III: Management style

The sample were particularly concerned about what management style to adopt in male-dominated organizations. The low-key style the majority favoured has certain disadvantages. Some people were exploring how to strengthen it by either using their femininity or adopting male tactics.

IV: The influence of work role on managers' private and social identities

How managers behaved at work had repercussions in their lives at home, some of which they saw as disadvantages of their jobs.

V: Managers' images of themselves in the future: being old means being hard

The managers could identify few women above them in organizations who they wanted to be like in the future, and this was a cause for concern.

In this chapter I am again borrowing language which does only partial justice to women's view of the world. 'Personal identity' tends to conjure up images of autonomy, individualism and strength through independence. Most of the managers did not subscribe to this view of being a person. I shall therefore use these words with caution, and move towards spinning new meanings for 'identity' which express women's values.

One aspect of the division of material between this chapter and the next, which deals with lifestyle, requires early mention. In general, identity (particularly in relation to lives outside work) was more a concern to single managers, and balancing different elements of their lifestyle to married managers. Both groups shared in all the dilemmas covered but to differing degrees. As I go through the two chapters, I shall indicate where these differences within the sample become important.

I: GENDER: IS BEING A WOMAN IMPORTANT?

As I set out to interview the thirty managers in the sample, I wanted to know how they saw the world and so used fairly general and open-ended questions. There was, however, one deliberate twist to my interview format, and that was to leave any mention *by me* of being a woman as potentially important until very late in the sequence of questions. I wanted to see whether *they* selected this as significant.

Our conversations justified my reservations. As a group the managers were strongly against identifying themselves as disadvantaged. There were a few exceptions who were keenly aware of difficulties associated with being women, but even they said 'I wouldn't have it any other way'. The managers mainly saw themselves as having done well, and lucky. 'Privileged' was the word some used.

But that word made me uneasy, and I realized that despite my espoused detachment, there was a sense in which I somehow disbelieved the speedy majority response that being a woman was not important. Intuition told me that this was an incomplete answer at best. Statistics revealed just how few women occupy management jobs. And no-one I discussed the matter with at the time would fully believe the managers' apparent opinion either. I soon realized, too, that the managers themselves contradicted this response in comments about gender and their awareness of it, and often passionate discussions of attitudes or behaviour they adopted to overcome the disadvantages of femaleness. These were clues that I should probe more deeply. As I drew together relevant elements, a more complex picture, with conflicts between different levels of expression, emerged.

At first I found myself in a dilemma of interpretation. The managers' opinions were quite straightforward and clearly expressed. They said they had not been, nor were, at a disadvantage because of being women. About half did mention their sex without prompting, but more did so to identify it as either an advantage or irrelevant, than as a disadvantage: 'I really don't feel that I've either got or failed to get anything because I'm a woman' and 'I don't feel prejudiced against, I know that's unpopular' are typical expressions of their views. Consistent with this were their attitudes to other, potentially less successful women. Whilst a few were sympathetic, more saw the Women's Movement as 'complaining too much', and several were extremely forceful in wanting to say 'it's not difficult' to other women:

'I suppose because I've got on and done it, I think anybody else can. I tend to be intolerant of people who moan and say "circumstances did it to me". You should get on and change them if you don't like them. And that applies to men and women equally really. We all have our problems but they are different.'

'I think there's a lot of rubbish talked about women being stopped from getting on. Women have to realize that if they do want to there's an awful lot of things they've got to take as disadvantages as well as advantages . . . Often they aren't prepared to accept what goes with responsibility.'

'If people say its difficult for a woman to get on in a particular company, I often take it with a pinch of salt, because I do believe that people can get on. The hurdles are there to be jumped over. If you're in a particular industry and can't get on you can always go to the opposition and do so.'

As I built up a sounder picture of the complexity and diversity of the managers' comments, I found I could accept these espoused views if I balanced them with alternative, often conflicting, material. I realized that this balancing act represented a process which many of the managers themselves performed, and discovered that one way of managing the potential stress of being a woman in a man's world was to control ones awareness of being female. I found I could subdivide the sample into three broad groups according to the kind of awareness they showed, and that this made sense of otherwise confusing material.

More than half the managers appeared to have little awareness of their gender, and said that being a woman was irrelevant. I have quoted many of their comments already as the majority view. But most held other opinions which contradicted these views. Three managers were acutely conscious when I met them that they were female and living in a male dominated world, and needed to cope with the frustrations and dilemmas this presented. Finally, a third group, about a third of the sample, accepted that as women they were at a potential disadvantage at work, and built their identity and approach to employment on this foundation. They were not experiencing the intense concern of the previous group, but their accounts suggested that several had already come to terms with their gender through similar experiences. Let me describe each viewpoint in more detail.

The majority of the sample held opposing views about womanhood at the same time if I take the full range of their relevant comments into account. They expressed overt opinions that being a woman was largely irrelevant to their experience of work, felt that both sexes have their different problems, and dismissed some which they had encountered as 'not very important'.

All also either identified disadvantages of being female, explained how they avoided or coped with discrimination, identified negative consequences of being working women for their personal and social lives, or believed that women in general are different from men and at a disadvantage in organizational life. A high proportion expected to meet obstacles to future career development because they were women. Many said explicitly that it is a man's world, and advised women to appreciate this in their approaches to employment. They went on to say that achieving a satisfactory work life is more difficult for married than single women, and even more difficult for those who have children.

I was able to understand how these apparently conflicting opinions fit together once I realized how this group had moderated the potential disadvantages of womanhood for themselves as individuals. Their main tactic had been to mute their own awareness of being women and different from men. They distinguished themselves as particular cases from women as a general group. They explained that they had not been aware of being female. By thinking of themselves as people or their professional role first, they minimized their own sensitivity, and thought they might also influence people with whom they dealt not to see sex as a relevant factor: 'I never emphasized being a woman, so I never made other people conscious either'. Managers were, in fact, irritated by times when attention had been drawn to their femaleness, as in the by-now familiar training course introduction: 'Good morning gentlemen . . . oh, and I see, *lady*!' They did not welcome the resulting visibility. Some managers explained how they minimized their contacts with other women at work to preserve their individualistic status, and avoid being seen as 'one of the girls'. Their immediate personal context had become all-important. As long as their private strategies for survival and achievement were effective, they had no incentive to identify with a larger grouping, and to risk the public visibility this would involve: 'This women's lib thing has never worried me because I feel liberated without having to fight. I have a very good relationship, I think it comes back to one's family, whether you feel your husband is holding you back or not.'

The tactic of muted awareness leaves some managers with a vague uneasiness that their femaleness may be significant to other people and a potential disadvantage. But this is an avenue of speculation they prefer not to explore. They say that they would rather not know as it is something they can do little about. Some also hinted at the difficulties they might face if they become more aware of being women, and expected that they would find these disabling. They might, for example, adopt a limited perspective and label all problems at work inappropriately as 'because I'm a woman'. Even if such an analysis was realistic, they felt they would be left not knowing how to behave, and robbed of responses they could previously have made automatically. The managers also feared losing their 'sense of proportion', an asset they prized highly. Above all they were concerned to avoid becoming angry. As one put it, 'There's a danger of becoming over-aggressive if you have too strong an idea of yourself as a woman fighting against men'.

Drawing together the various strands, it seems to me that this substantial group within the sample was allowing that there are potential disadvantages to being women, but they believed that awareness increased the likelihood of these becoming significant. They were happier to disregard these issues in their surface consciousness.

The three managers who spoke at length about the problems of being women illustrate the painful turmoil to which greater awareness can lead. For all three, recent circumstances had prompted an acute consciousness of themselves as women, which had shattered the relative calm of their previous perspectives. One was experimenting with a new, more aggressive management style in order to survive amidst the cut and thrust of senior organizational levels. Ironically, she felt colleagues' reactions against equal opportunities' publicity were making life more difficult and confronting for her. The second was dissatisfied at work and trying to decide whether she was in the right occupation and company. She saw her employer as favouring male stereotypes of management, and unable to recognize and accept her different style of working. The third wanted to combine having children with a career. Her sense of herself as female had been heightened by the obstacles and difficult choices she faced.

All three managers were finding this a stressful and relatively isolated time. Their attitudes and feelings were shifting radically as they adopted new perspectives on the world. One felt she had previously been 'blind to what has truly been happening'. They became particularly sensitive to apparent prejudice and discrimination. How to be and to behave, especially in close relationships, had become problems. They felt they were often behaving inappropriately, and taking things too seriously, although other people did not always seem to notice. When I met them none had yet evolved a new set of values or perspectives which acknowledged their awareness of gender and welcomed it as acceptable and strong. The attitudes of the third group on the continuum of awareness suggest that this will eventually happen. As there is no going back from the conclusions they have reached (Daly, 1979), they must go forward.

The third group of managers had already developed a clear sense of themselves as women. Work experiences seem to have at least contributed to this. Most of these managers were from general management or (retail) buying jobs, or had been involved in setting up union branches in publishing companies. In these non-traditional jobs or activities they had faced a more explicitly male-oriented culture; had compared themselves with male colleagues and identified inequalities in reward; and many had been openly told at one time or another that there were things they could not do because they were women. That they were different and treated differently seem to have been truths they could hardly avoid. These managers showed greater concern about the position of women as a group, and of themselves as women, than did the majority of interviewees. They brought analyses along these lines into their answers, but seemed personally untroubled by them. Some translated these attitudes into an advocacy of women's rights. Others took up the possibility of discrimination against them as a challenging gauntlet. Several of the latter expressed an edge of bitterness or resentment, but this was typically balanced by cynicism and self interest. 'But I wouldn't change it, its exciting and challenging' was a typical comment.

Through the analysis described above, I discovered how the managers varied in their appreciation of gender and their inclusion of being women into a broader sense of identity. The data also chart a possible sequence of increasing awareness which some women follow as they integrate being female more fully into their sense of themselves in relation to others. For most women any such journey will initially involve a painful confrontation with a new way of looking at the world, an acknowledgement of men's domination of society and their power to constrain female development. Growing through this realization to new truths of their own is both a challenging and dangerous opportunity. It is one increasing numbers of women are taking as social and media attention to women's issues prompts them to a new appreciation of gender. The alternative, of avoiding awareness, is protective but has its costs. Central areas of the manager's being are partitioned off, unable to contribute to her full capability and identity. Hennig and Jardism's (1978) sample's need to pause in mid-career and redress their previous neglect of themselves as women conjures a powerful image of necessity about the task of reintegration. Their managers took time out to develop their female characteristics, but individuals will vary in which areas of self they need to explore. Some women (particularly those who work in caring occupations) may need to honour the agentic aspects of their self, whilst others will need to allow muted communion expression.

The chapter's next main section fills out an aspect of the data I have already touched on, the everyday problems the managers experienced at work because they were women, and their strategies for coping with them.

154

II: EVERYDAY PROBLEMS OF BEING WOMEN, AND STRATEGIES FOR MANAGING THEM

As I have already indicated, the managers were ambivalent about how significant the day-to-day problems of being women were to them. They were often tempted to dismiss them verbally as 'not very important', and to balance their talk of problems by identifying advantages of being women. These aspects of working were, however, more significant than this apparent dismissiveness suggests. The managers talked at some length and with feeling about the problems they encountered and the strategies through which they tried to manage them. They gave frequent examples, especially if they were perplexed at how to handle the situation better, or proud of having done so skilfully. I have already met this phenomenon in Chapter 4 in the discussion of the consequences for women of their exclusion from informal organizational networks. Particular examples seemed somehow 'trivial' and 'petty'. Their real significance comes from the total picture of inequality to which they sum, and which they express. It is part of that picture that women do not think they have the right to complain about how they are treated by others.

In this section I shall move rapidly through a list of the problems of being women that managers identify, and then a list of their coping strategies. These show the spread of their concerns, and illustrate the devaluing of the female identified in Chapters 2–4 in practice as constraints on women's behaviour in management jobs. I shall not elaborate further on their meaning. The underlying issues expressed tended to be vague, insidious, deep-rooted, constantly present or potentially so, pervasive, almost always open to alternative interpretation, and above all difficult to address. As one might expect their main subject matter was difficulties in relationships with men at work. These lists reveal the options managers perceived and the values they used in choosing between them. They illuminate the context within which managers weigh options of management style, and suggest repercussions of different choices.

Problems

Not being taken seriously; feeling patronized Not being taken seriously was a common concern, but not one managers could always back up in substantive ways. I shall illustrate with one of the more identifiable examples:

'I've had a couple of occasions when men have obviously tried to side-track me. You come in talking about something very important and they say, "Oh, you do have lovely eyes". I had one man who did this all the time. Once he came in very annoyed with me about something I had done and I said, "Mike, what a fantastic tie you've got on". That was the end of that and he never did it again, and he was quite aware of what he had been doing.'

More often managers were left with the vague feeling this might be happening, but little firm evidence to support their suspicions.

Feeling misunderstood Some managers thought that men at work had false impressions of them as people, and did not really understand what they were like and what motivated them. Some of the misconceptions more commonly suggested were that working women must have unsatisfactory marriages, could not possibly be feminine, neglect their children, cannot be worth the money they are paid and are unlikely to take the job 'really' seriously.

Being wrongly identified, underestimated Managers were often mistaken for secretaries or assistants, particularly once they moved outside their own companies in which people had come to recognize and accept their status. I can vouch for how natural others' (mistaken) expectations usually seem to them at the time. Waiting coldly outside a station while a driver searched for a male 'Doctor' (and speaker) to deliver to a waiting audience is one of the less consequential examples from my own experience of this happening.

Being judged on double standards Many managers felt that similar behaviour is judged differently according to whether it is displayed by a man or a woman. If a woman shows anger, for example, she is likely to be seen as 'emotional' and 'hysterical', whilst a man behaving similarly is 'just letting off steam'.

Underachieving because of others' low expectations It was generally agreed that people do not expect as much from women as from men, and that women themselves can easily come to believe and act from these low expectations.

Managers believed that most men find it difficult to accept instructions and authority from women; and men who work for or with women are laughed at by their wives and friends Some interviewees felt that, as a result, they must contribute to the humour about the unusualness of their authoritative position.

Facing choices about what to wear This is where one of my prejudices must be admitted. Dress is a topic I really did not expect to discuss much, as I see it as unrelated to getting the job done, competence or promotion potential, and feel insulted that in general conversation about women and work people seem to take so much notice of incidentals. But I have had to drop this idealism and bow to the weight of opinion. I now include dress as a factor influencing women's experiences of employment here with true conviction. Managers themselves mentioned the significance of clothes frequently in their accounts and explanations, and I was interested recently to hear an all-male group discussing women at work devote much of their time to it too.

So what importance does dress have? The women managers' main complaint was having no uniform (like a suit) to reduce the burden of choice of what to wear at work. Having to find the time, interest and money to pay attention to their clothes and general physical appearance was also a concern. The male managers (whose group I observed) envied women their potential variety in clothes and felt they were stuck with dull uniformity in comparison. Some also felt that women make life difficult for them (men) if they dress in ways which heighten their sexual attractiveness.

Norms about dress reduce uncertainty by helping organization members identify and classify each other and enabling individuals to signal their appropriate organizational slot. As yet, there are few established standards for female dress which can serve such functions, although some conventions do seem to be emerging — the neat suit, with its clear parallels to male norms of dress, which I now meet in many executive offices, for example. The norms of female dress we do have are derived from contexts other than employment, and echo the patterns of male–female relationships characteristic of society. Their available categories will tend to approximate the stereotypic female role traps identified in Chapter 4 of mother, seductress, pet and iron maiden. Uncertainty about working women increases the likelihood that symbolic categories such as these will be imported into organizations, and clothes used as clues by which to classify individuals. Dress can therefore serve symbolic functions, moving individuals towards or away from different categories. The choices women make about what to wear become understandably important when viewed in this context.

Whether women pay attention to their appearance also has an interpretation value well in excess of its apparent significance. If she does not actively take trouble over how she looks, the manager contravenes basic expectations of femininity, and again raises others' anxiety about who she is.

Strategies

Managers told me how they did, or thought they should, cope with the prejudices against women identified above. They also offered their strategies as advice to young women entering employment. Their most common suggestions are described below. This is not a comprehensive list, but gives a flavour of their approach. A theme which links many of their strategies is one I return to in other contexts — that of feeling personally responsible. The managers often appeared to be acting to protect the interests and feelings of others, at their own expense. This was a strong, almost automatic, reaction. Later in the chapter, I report that some interviewees are now deliberately trying to unthink this response, and to take care of themselves first.

Don't look for prejudice As the section on awareness of gender above has indicated, the most common strategy for dealing with peoples' attitudes to women at work, was not to look for prejudice. The manager would then be able to behave 'naturally', would not be faced with the problem of what to do and felt that others would be less likely to regard her sex as significant. Their advice to others was to be a person and an individual first, rather than a woman.

Do the job well Also popular was the strategy of doing the job so well that one is above criticism:

> 'The best way to get over prejudice is just to demonstrate you're as good as the next man' . . . 'I think if you are very efficient and do the job very well and you are also reasonably pleasant to get on with, it would be a foolish man who would go out of his way not to have you as a member of his staff.'

(Managers often noted how good they were as assistants, and concluded that others could hardly deny them opportunities as they made such valuable subordinates. The people who made the decisions were perceived as male, and managers often seemed to rely on their perceptiveness and good faith in recognizing and developing talent.)

Get appropriate qualifications

Build up credibility, gradually Many interviewees took it for granted that they would initially be under-valued by those they met. Their typical reaction was to let people get to know them better, rather than try any deliberate tactic to display their competence. They identified staying with one company as helpful in minimizing problems of credibility.

Be persistent, and confident Persistence was highly prized, and seen as likely to lead to recognition and further job opportunities. Maintaining at least an outward confidence is also necessary in order to behave as they recommend:

> 'Don't back down just because you're wearing a skirt. If you stick at something things come your way. If you want to run things don't imagine anything will stop you. You can do anything if you're determined enough'.

> 'I think persistence is probably the key word. Make yourself felt and heard. Some day someone will turn round and say, "Oh, hasn't so-and-so always been going on about wanting to do that? I'll give her a chance".'

The manager should back up persistence with her own belief that she is capable.

Take care not to fulfil negative stereotypes of women 'Women have to be much more careful about how they behave and are seen to behave,' I was told. I was offered lots of advice on what to avoid. Being aggressive or emotional, losing my temper, shouting and dithering were common examples.

Adopting a non-competitive style The managers were very concerned about what style to adopt at work. Most favoured a 'low key', 'subtle' approach as being compatible with their self-perceptions and effective in the face of male colleagues. The need to choose, and the dilemmas it raised, were so important to them that I have given them separate consideration in this chapter's next main section.

Adopting a non-competitive style as a reaction to prejudice in fact reinforces a traditional pattern of relationships in which men are dominant and women adopt more subversive approaches. As I have concluded before, a coping strategy which protects the individual can actually block longer term or more general changes which would make women acceptable as themselves within organizations.

Unexpected confrontation Some managers occasionally supplemented their subtle style with a more confronting approach. They felt this is all the more effective as it is unexpected, and men find it difficult to confront women back.

Deliberately 'un-think' low expectations copied from others Some managers reported times when they had suddenly recognized that they were letting others' low expectations influence their sense of competence. They would then deliberately unthink these attitudes, telling themselves instead that they *were* capable.

Put the problem into perspective as 'just another of life's problems' 'All the time you are in management you are facing difficult, complicated situations. This is just another one that you have to learn to tackle'.

Keep it lighthearted, and a subject for jokes Some interviewees felt it was their responsibility to maintain a sense of balance at work, and would encourage or make jokes about their position if this relieved others' tension about accepting them as managers.

Ensure job title and role signs do adequate justice to ones status Several managers explained how they paid attention to significant cues to ensure that their self-presentation left as little room as possible for others to misjudge their status. Some, for example, spent a lot of money on clothes for work with a deliberate eye to their visibility.

Be prepared to confront prejudice directly Some managers said that ultimately one should be prepared to confront prejudice openly if other tactics were ineffective.

Anticipate and plan for potential prejudice Despite their own majority pre-ference for subduing awareness of potential prejudice against women, the managers enthusiastically advised others to take a more penetrating view before embarking on employment. The planning they advocated had three aspects. The first was knowing what one wants in terms of level of commitment to work, career aspirations, marriage and having children. Secondly, young women should anticipate potential problems and develop strategies appropriately. Finally they should critically evaluate the profession and organization they plan to enter. Having assessed how welcoming to women both are, they can judge whether the level of challenge suits their motivation and abilities. 'Not everyone has to be a pioneer'.

Actively search for role models Managers advised younger women to look out for females who are successful in ways they aspire to, and to study how they behave.

The managers' list of 'problems' reveals how devalued by others at work they could feel. From their list of strategies we see how they often took on personal responsibility for managing the disturbance they represented in the work environment. They tended to handle potential prejudice by accommodation wherever possible, adjusting their own attitudes and behaviour to counter contextual pressures. They would then pursue their own purposes by covert rather than overt means. Only occasionally did they openly challenge the environment and the expectations it held of them. Their approach is highly reminiscent of the typical pattern of female–male conversation illustrated in Chapter 3, with women oiling the wheels and handling uncertainty and risk for the system as a whole. Within this context of pressures, personal motivations and anticipated consequences, women choose their management style.

III: MANAGEMENT STYLE

Many managers expressed their concerns about identity as the dilemma of what style to adopt in day-to-day relationships at work. The choices they faced were not simply issues of personal preference or ability, but tackled head-on the operation of sex-role stereotypes within organizations. Interviewees themselves brought interpretations of stereotypes to the debate. They identified ways in which women and men are different as the foundation for their discussion of styles and options. Whilst they covered a broad range in terms of their own personal characteristics, the sample were highly consistent in describing a core style which they felt builds on qualities and *advantages* they have as a result of being women. They showed considerable commitment to its characteristics, its compatibility with their values, and to the contribution it can make to organizational life. Their statements are an affirmation of women's experience.

The managers did, however, see some limitations to this style's effectiveness, particularly at senior organization levels. These were prompting some to consider adopting alternatives. They saw two broad choices: 'using ones femininity' and 'adopting male tactics'. These are virtually caricatures of traditional sex stereotypes. Some managers have found elements of them useful, and are now grafting these on to their basic style. The discussion below follows this sequence of development: a description of the sample's core management style is followed by an identification of its limitations and, later, considerations of alternatives.

Elements of a woman-centred management style

The core style managers favoured was underpinned by three main themes:

(1) a belief that women have access to more varied and what they called 'softer' techniques in personal relationships;
(2) their understanding of, and sympathy for, others' needs, particularly those of subordinates';
(3) women's greater freedom of choice in relation to employment.

Their approach to management was based on principles of cooperation rather than competition, and so confirms this book's earlier, theoretical discussions. This becomes particularly apparent in the values which inform their preferences and choices. They believed in honesty, authenticity and cooperation, and consistently rejected superficiality, putting on false appearances and aggressiveness.

Women's softer techniques

The sample firmly believed that women have a broader range of possible techniques for relating to others than do men, opening up alternative ways of communicating and getting things done. Listening before acting; being sympathetic; being trustworthy; not needing to dominate others; getting things done by being nice to people; making friends with people so they try not to let you down; being prepared to apologize if something goes wrong; being able to say personal or threatening things to others which would be unacceptable coming from male colleagues; these were some of the many illustrations given. As one manager put it, 'you make it a gentler atmosphere in some way'.

Managers' preferences for honesty and openness, and if necessary bluntness, are particularly apparent in these comments. In this sense it seems that women bring their traditional domain, the private or personal, into the workplace. Wherever possible they establish it as a basis for getting things done which suits their personal style and skills. Being content to build up credibility by letting people get to know them is a similarly founded strategy. Their tendency to work through private strategies wherever possible helps explain why public-oriented activities and the workings of organizational politics at senior levels came as a shock to many women who had been highly effective lower down the organization. They had been doing jobs by different methods, and had not necessarily developed the skills, or perhaps more importantly the associated values, to copy these aspects of male colleagues' tactics.

Understanding others' viewpoints

Interviewees felt their powers of listening and empathizing helped them understand others' perspectives and improved their effectiveness as managers. Many also still remembered occupying lowlier organization positions as additional information on subordinates' likely opinions and feelings. Some of the specific ways they translated these understandings into action were showing respect and tolerance generally, delegating work which challenged subordinates' capabilities, and encouraging participation from others. Again, their emphasis was on fostering cooperative relations as means to achieving more specific work goals.

Perceived freedom of choice

Faced with a wider range of options than male colleagues of how to spend their lives and, for various reasons, often being less concerned about financial security, most managers felt they had greater freedom of choice than do men. This is demonstrated in their insistence that work is only one of several important life areas. They were therefore less concerned about excelling as they did not rely on work for their sense of self-worth, and many were so lightly attached to a particular career direction that they were willing to consider alternatives and make radical changes. In particularly situations they felt this independence of work or employer lead them to take risks whilst they saw colleagues hold back. Being prepared to ask awkward questions in meetings with little concern about whether they appear stupid was frequently given as an example.

The translation of the interviewees' preferred management style, which they termed 'low key' or 'subtle', into action means getting things done by influencing others rather than direct control, little concern about publicly taking credit, founding business on amicable personal relations, cooperation rather than competition, getting on with tasks and paying little attention to appearances. At its best this style can be highly effective, and foster a working culture in which others participate similarly. This was reported to happen in certain areas of publishing where concentrations of women are high, and was sufficiently marked to be noticeable to outsiders:

'Someone came to a function here and said afterwards that she had never been in a room with so many high-powered women who didn't compete with each other. We'd never thought of ourselves as a group of competitors and it was really creepy, it put a completely different slant on it.'

In hostile environments such strategies could leave the manager bruised, battered, unable to perform effectively and feeling undervalued. Several managers who were re-evaluating their approach had been prompted to do so by experiencing stresses of this sort. More routinely there seem to be limitations to the applicability of this style in organizations as they currently function. Within the sample there was a difference between the two industries in this respect. Publishing was more accepting of creativity and working through relationships, and valued managers of either sex for these characteristics. Retailing was more polarized in its stereotypes of appropriate behaviour. The low-key management style described was not generally approved of. Even in staff management jobs, employees were expected to show a commercial edge in their approach.

By the time I met them many managers from both industries had encountered situations they could not handle satisfactorily with their preferred style. Some had also concluded that their career opportunities were limited because their talents were invisible to evaluation systems and personnel in the organization. These various difficulties had prompted them to look outside their natural base for alternative strategies. The two main options they identified were 'using ones femininity' and 'adopting male tactics'. The approaches these labels refer to are not radically new; in fact they group together relatively stereotyped ways of behaving. The issue for the sample was more whether to abandon their basic philosophy of behaving naturally, and to borrow from either extreme deliberately, artificially. In their evaluations of the various techniques, four considerations were important: authenticity, gender stereotypes, job effectiveness and personal survival.

Authenticity was a key issue. The managers wanted to behave naturally and not to put on a performance, which they saw as being inauthentic and thus manipulative. A manager talking about the woman-centred management techniques described above illustrates the force with which such comments were usually made: 'I think its bad practice to do that artificially. I wouldn't approve of that being done consciously'. Alternative strategies for improving their management style were therefore acceptable only if they could adopt them artlessly.

This basic philosophy's expression of women's characteristics has significant implications for their behaviour as managers, and their development and learning. The managers' insistence on authenticity originates in notions of communion rather than agency. The communion-based talents women managers identified as their own, cannot easily be developed and improved deliberately as this contradicts their assumptions of openness, instantaneous relevance, flexibility and change, and the values in which they are based. Male techniques which are control-oriented can be learnt and developed without this complication. In this sense there seems, then, to be an implicit limitation to women managers' potential to develop their personal base of talents, unless they move towards strategies of communion enhanced by agency. The necessary learning need not be a deliberate, concept-based process. The managers' accounts illustrated alternative strategies of development through exposure to challenging situations or events which had prompted radical revisions of their perspectives on work. Learning thus emerged from being and doing, rather than originating in prefigured plans.

A second criterion by which managers judged alternative ways of behaving was compatibility with their gender-based self-images. Most could not reconcile description such as 'aggressive' or 'ruthless' with their views of themselves. Many, however, lived with some conflicts in these terms and behaved in ways which put their more traditionally female self-images at risk.

The two remaining criteria for evaluating alternative strategies — job effectiveness and personal survival — are often in direct conflict with authenticity and gender images. Managers placed a high priority on getting the job done well. They were therefore sometimes highly disappointed to find that their reluctance to pay attention to what they considered non-essential aspects of organizational life, such as politics, had undermined their task performance. An initially unpalateable style might therefore be acceptable if it protected their more directly job-related efforts. Concern about personal survival might also encourage them towards compromise. Several managers had recently resolved to protect themselves more from undue pressure from others, or self-generated concern about things to do with work, especially if these translated into physical symptoms of stress. They talked of 'learning not to mind'.

Having achieved some understanding of what managers were seeking and avoiding, I shall now look in more detail at the two broad alternative approaches they had identified as possible supplements for their original, sometimes vulnerable, low-key management style.

Using ones femininity

Whether or not, and how, to 'use' ones femininity was a major concern. This had a range of interpretations based on different individuals' stereotypes of female behaviour. Playing the dumb blond so that others (men) would offer assistance, trading on sexual attractiveness, and flattering male vanity were offered as examples. A more recent option was whether to take advantage of offers to be a token woman. Capitalizing, deliberately, on being a woman was generally, although not universally, frowned on. 'I would despise using being a woman to shove off responsibility, that I would think to be absolutely abominable. I might swear, but I wouldn't burst into tears', was a typical condemnation.

A significant minority, however, said they had overcome their initial 'scruples' (their word), and learned that in order to survive at work they should deploy whatever advantages they had. Several have struggled with the morality of this decision, and it is interesting to see the conclusions they had reached. They justified them as 'only fair' in order to balance their disadvantages and the advantages they believed men to have:

> 'I started off very much believing that if you were going to do the same job as a man then you would have to be prepared to take it on equal terms. But I have changed my mind altogether. I decided that men take advantage of the advantages they've got and I'm going to take advantage of the one's I've got. If I can find my way round things by being a little bit devious at times, I'm afraid I will do, now.'

'If I know I'm using a particular approach with someone that is allowing my femaleness to be the main influence, but this business issue will be resolved as a result, I am perfectly prepared to do that now.'

'I'm certainly not beyond using a bit of feminine charm if it will achieve the right objective in the end. So I would say my methods are different, but I don't think that's unfair.'

The full delicacy of when it is acceptable to use being female is reflected in one interviewee's advice to other women: 'Use your femininity if you can do it naturally rather than falsely, in situations where it helps. I don't think its a good idea to be aggressive about it, but don't feel you should hide your femininity either. I don't think it works if you try very hard to be a chap when you're not.'

Taking a longer term view, I found a similar balance of opinions. The majority of the sample wanted advancement only if it reflected recognition and appreciation of their abilities. Again a determined minority were more concerned about the ends and willing, if they must, to be 'philosophical' (their word) about the means. One woman in retailing, for example, anticipated that her company might think it 'politic', given equal opportunities legislation and publicity, to promote a woman to a 'very senior' organizational level. She was willing to take advantage of any such offer without major moral scruples:

'I might get on because they think it is "politic" and "wouldn't it be nice". I think, "you fools!" But if I can get there that way, then fair enough! My career is absolutely dictated by men, not one woman has any say in my progress.'

Adopting male tactics

The sample expressed even more reservations about adopting what they termed 'male tactics'. These were, however, typically seen as essential to effectiveness and survival at senior management levels. There are three characteristics of male tactics as they described them: aggression, paying attention to appearances and playing political games.

Managers varied in how compatible they thought the aggressive, confronting and competitive behaviour they attribute to men was with their own personal style and self-image. Some, particularly amongst those in retailing, felt comfortable with a more assertive approach. For the majority, however, this was totally unacceptable, and not something they wished to consider:

'I don't like aggressive women and steer well-clear.' . . . 'There is no need to be aggressive. I think its very distressing to see.' . . . 'I think women have degraded themselves no end (by being aggressive).' . . . 'I think its partly their (women's) own fault that they don't move up, they're less ruthless in many ways. But then that's the nature of women, and they're probably a lot happier for it really.' ... 'Its odd that women should go to extremes to emulate the men they say are inferior.'

Some younger managers felt that previous generations of women have had to copy men's aggressiveness to achieve acceptance, but that this was no longer necessary or desirable.

Several managers in retailing who were experimenting with different styles provided interesting illustrations of what happens when women behave more aggressively. They had decided, most reluctantly, that they had to take a tougher approach in order to survive and be effective at senior organization levels. One was noting with interest the effects this had on others and on herself. Her conclusions are typical. Other people did not appear to notice:

'I have had to try to be a bit more detached and, even when I don't feel it, to pretend that is how I feel. Probably that is a bit irritating because I am not behaving according to their idea of a female . . . I have started deliberately being a bit more aggressive towards them, and if somebody is deliberately awkward or rude I react in kind . . . It's self protection. It's a bit upsetting at the time but they don't notice anything different, that is the surprising thing. So I am gaining a bit.'

Instead, the repercussions have been in the conflicts between her new behaviour and her self-image. She initially felt upset at contradicting her perception of herself as someone who should help reconcile rather than add to interpersonal conflicts: 'I tended to regard it as part of my responsibility as being to take the steam out of situations somehow, but then you are a one-man band.'

A second male characteristic on which managers commented was that of paying attention to appearances, of being superficial. This often resulted in 'shouting the odds' at meetings in order to gain attention when they have little 'really' to say. The sample disapproved of such show; they described themselves as blunt and task-oriented in comparison:

'Something men have that I despise, but is very useful in internal politics, is the capacity to dress up 'I want' with a load of irrelevant verbiage. I'm very much closer to saying "I want" and that is very tactless, and not very wise.'

This manager thinks her own approach unwise because she transgresses the rules of how to get people to do things, and, she thinks, limits her chances of being promoted. Women's refusal to subscribe to conventions which sustain idealized appearances have earned them the reputation of 'demystifying' work, of revealing it as less difficult, less intrinsically meaningful than appearances imply. This cluster was, then, the set of 'male tactics' the managers seemed least disposed to adopt.

Thirdly, the sample saw their male colleagues as likely to play politics and to exercise power for the sake of doing so; this was sometimes described as 'playing games'. Much recent management literature encourages us to understand, and accept, the workings of politics within organizations. The impact of this 'unofficial' power system cannot be underestimated. Writers differ, however, on how 'real' or essential its workings are. Pym (1980), for example, has consistently condemned these as rituals of employment which he distinguishes from 'real' work. One reason why his views have had little identifiable impact on organizational life is that these rituals perform important functions, but these are emotional rather than task-related. Chapter 4 showed how informal systems handle uncertainty, risk and conflict, and protect the individual from the weight of full personal responsibility. Organizations can, then, be seen as mechanisms by which their members manage significant threats. As men have been the main shapers of organizational life I see its structures, procedures and politics as significant expressions of men's emotionality. The underlying concerns organizations are deployed to conceal and alleviate are therefore more likely to reflect men's traditional societal position and experiences, which will be significantly different from those of women. Organizational politics may therefore make less sense to women as they do not share the base of experiential sense, especially of threat, on which it is founded. To them it is a 'game' as they do not share the 'rules'. Perhaps, too, women and men tend to devalue each others' emotionality.

The managers in the sample tended to sit on the sidelines of organizational politics if they paid it any attention at all. Some had become fascinated by the covert workings of power, but only a few hinted that they also took part themselves.

Reflections

The managers' consideration of management style reveals the creativity required if they are to be effective, survive *and* achieve an identity at work compatible with their own sense of self. They see themselves as different: different from the male culture in which they operate, and also from women as a general class with whom to identify. Traditional stereotypes of what it is to be female also haunt their discussions, both as contributions to their own self-images and powerful forces shaping other peoples' expectations and reactions. The managers' commitment to cooperatively attuned, communion-based values and ways of being and doing, can leave them vulnerable in this discordant context. Their responses vary. Those who plan to persist with management jobs seem mainly to be moving towards enhancing communion with more agentic tactics based in confrontation and assertiveness.

IV: THE INFLUENCE OF WORK ROLE ON MANAGERS' PRIVATE AND SOCIAL IDENTITIES

Who to be was not only an issue for the managers in relation to work. All depart from traditional notions of womanhood in their lifestyles generally, and are therefore faced with choices and dilemmas about identity in their private and social lives too. As employment and the type of job they did were such central factors for them, their concerns were expressed in terms of the repercussions these had for other aspects of their identity. Working well above organizational levels typical for women, the nature of the industry and their particular job functions were separable, but interacting, influences. The impacts they identified and how troubled they were by them varied greatly across the sample. Younger, single interviewees were most keenly aware of potentially negative repercussions of working. There were also significant differences between publishing and retailing. I shall first set the scene by looking at each industry separately, and then consider in more detail the dilemmas young, single managers faced. This section closes with some impressions of how the interviewees handle issues about social identity.

Industry and job function impacts on social identity

Publishing was frequently described as a job that takes over ones whole life. Needing to read a lot, the concentration of the industry in London, a busy social life through which much business is conducted, as well as high levels of involvement from the sample, contribute to this. People in publishing tended to work long hours and socialize only with others in the industry. The disadvantage of this close, busy lifestyle were feeling that the job was taking them over, having little time to relax and unwind, and making few friends outside the industry. Its advantages were, however, also substantial. It provided a ready-made pool of like-minded people as potential friends. As women are accepted in the social network, the pressure some working women report of being excluded from informal decision-making was not an issue.

The repercussions of work on their private and social lives were not, on balance, major concerns for most of the managers in publishing. Only younger, single interviewees were significantly troubled, regretting particularly that they had few friends outside the industry. This small group identified difficulties which their contemporaries in retailing voiced more vigorously. I shall return to these below.

Working in retailing was also characterized by high commitment and time demands. This industry, however, differs from publishing in several ways which put managers' lives outside work under greater pressure. The skills which are traditionally valued in employees involve being competitive and assertive, and therefore lead managers further away than publishing does from socially acceptable female behaviour. The industry's geographic dispersal and small management teams mean that interviewees were unlikely to have a ready-made circle of friends based on work to mix with. Regular geographic moves also confounded people's efforts to establish viable social lives. Most of the difficulties interviewees in this industry identified concerned the resulting sharp divide between their work and private lives. This had different impacts depending on the type of work they did. The jobs done by interviewees in retailing appeared selectively to accentuate aspects of their identity in ways which then inhibited or shaped their private and social lives. The impacts were different for the two broad occupational groups within the industry, staff and commercial management.

Many of the interviewees in staff management did not want to mix socially outside work because their day had already been taken up with being sociable and dealing with other people's problems. Married women and their husbands tended, as a result, to lead very self-contained lifestyles, into which only a few close friends and members of the family were invited. Single women did not feel that they could allow themselves this option, and expected a lonely old age if they did so. Many, therefore, made an effort to maintain a social life.

Women in commercial and general management functions experienced a different impact of work role on identity. Most found that the kind of person they were at work — competent, professional, business-like, unsentimental and managing — was not 'viable' as a social self. This profile was not acceptable to other people and conflicted with their own images of womanhood. Both interviewees' and others' discomfort originates in their, at least partial, acceptance of traditional sex role stereotypes which women managers inherently contradict. The sample thought that people with whom they mixed socially were intimidated by their abilities, and appeared afraid themselves of the power they had to dominate others if not controlled:

'I want to be dominated, I suppose because if I'm not I will dominate them. It's what I'm trained to do all day — telling people what to do, making decisions and planning things. I find with relationships outside, if I don't have someone who will do that I will take over and in the end that ruins the relationship, and I'm in charge and that is not the way things are designed to be.'

As this reasoning illustrates, part of the female managers' dilemma stems from the narrow range of qualities social norms allow each sex to display.

Some managers accommodate to potential conflicts by being different people in their two worlds — at work the leader, socially the led. Others were uncomfortable at fragmenting their identity in this way and felt that they were denied as whole people in the world outside work. One married woman illustrated this conflict by her dislike of mixing with other 'wives' at the golf course, and having to put up with 'strange attitudes' from her husband's friends. These were sometimes expressed in snide remarks about whether she could possibly deserve her salary or cope with the job, for example. Another spoke of the strain tension about identity can put on working women's marriages. I was sometimes told that divorce rates amongst upwardly mobile women managers are high, but was unable to confirm this. Only three of the thirty people I interviewed were divorced or separated.

Young, single managers: distinctive problems and opportunities

In both industries, managers with long-term partners tended to be more satisfied with their identity outside work than were their single colleagues. The former stated firmly that their marriages were founded on the fact that they were employed. This corresponded with their own and their partners' expectations and wishes. They therefore had at least one further arena in which their work identity was largely accepted and affirmed. Amongst single managers, attitudes on this issue varied, chiefly with age. Older women were more likely to prize their independence. The younger managers, on whom I shall now focus, tended to be very aware and concerned about their potential social isolation. Most of the young, single managers in the sample not only held fragmented views of their identity, but also saw their work roles as the 'real' them. Their lack of confidence outside work tempted them to persist in this relatively safe identity, and they found mixing socially a strain:

'I think work is the real me. I have to work harder socially . . . I have to try to develop the other side and be normal and really turn off.'

'In business I find if you do the work you are accepted as a woman and as a person; but I find in society you are not accepted as a woman in business, and that's the problem. You can't prove to society that you are a good manager or you're not butch and you are quite a normal female, but you just happen to be in a management position rather than a secretary or someone's assistant . . . It gets to the point where I feel like saying I'm a secretary at the BBC. Painful things like the car I drive. I would rather it was a mini than what I have, the reactions I get all the time get on my nerves. Then I have to justify how I have become what I have. It doesn't happen to men, they don't have to explain why they are marketing executives or whatever, or talk about their work at every dinner party they go to . . . I have quite a few girlfriends. We prefer our own society sometimes because we don't have to go into this sort of rubbish (explaining who we are) every five minutes.'

The managers' general difficulties of social unease and inability to find an identity which was both acceptable to others and compatible with their views of themselves, were accentuated in intimate relationships, particularly with potential sexual partners. Most of the younger, single managers expressed some dissatisfaction with this area of their lives.

A complex of factors contributed to the managers' uncertainties about close personal relationships. The identity they presented socially was often deviant and traditionally unacceptable for a woman, leading to various forms of rejection and aggression from other people. The managers themselves may have contributed to confusion about who they were by revealing and playing out in their relationships conflicts between various aspects of their identity. Several spoke of oscillating between competence and their co-existing need to be dependent, in ways which unsettled other people. There were the more practical problems that social norms frown on women moving around freely alone. They were unlikely, for example, to be invited to dinner parties unless they could muster a partner. Who to invite and how to pitch the resulting relationship highlighted dilemmas about their solo status. Also, the pressures of their life style limited their opportunities to meet new people and meant they had little time to nurture new relationships. Several of the publishing managers keenly wanted to make new friends, and for these to be outside the industry. Ironically, whilst publishing's informal networks accept women as people, they can also constrain them from expressing aspects of identity which are not demanded by their work roles.

The sample met few men they considered both suitable and capable of relating to them as women who were also competent managers. Several were acutely aware that the pool of potential partners is limited, and likely to decrease as they become more successful at work. The managers' specifications for an ideal boyfriend were exacting, and included several potentially conflicting elements. Most agreed that they were looking for someone who was worthy of respect, successful and confident in their own right; would not be threatened, overwhelmed or dominated by a successful woman; was dominant and protective; was able to be proud of them; and would be prepared to share in running a home. Their accounts of this ideal, and of relationships which had gone wrong for them in the past, are strongly evocative of traditional stereotypes of feminine and masculine characteristics and behaviour. In many senses they are caricatures of such profiles. The managers wanted to meet people who were highly masculine in traditional terms and could keep their own developing assertiveness in check. To complement this they were prepared to display highly feminine elements in their own personalities. They seemed to want to be female in ways their jobs did not encourage or allow.

Even once they did meet potentially suitable people, the managers doubted whether their lifestyle would allow them to develop the relationship. Geographical moves, frequent in retailing but also possible in some branches of publishing, could disrupt (for some had already disrupted) friendships. Some managers admitted, however, that they found this intrusion helpful in a way. It protected their uncertainty about who they were socially by allowing them to keep relationships on a superficial, and therefore more manageable, plane. The pace of their lives also meant the managers had little time to nurture developing relationships in ways they saw as part of getting to know someone well:

> 'One of the great dangers of working as hard as I do is that you have less time to try, to make gestures. Like thinking if someone's coming it would be nice to get some flowers because it would be nice . . . ha! When did I last get some flowers because that would be nice?! Years ago . . . and that's a pity, I think.'

Given problems of finding suitable partners and of how to present themselves socially, several managers described how selective they were in the company they kept. They preferred to mix with people who accepted them in their totality, without thinking them odd, and to whom they did not have continually to explain themselves. Some, more from publishing, emphasized the importance of maintaining links with old friends and the strong sense of identity and continuity this gave them. Often this also lead them into god-mother relationships, which they found highly satisfying. They had invested this role with new meaning as one for extending nurturing and affection outside the narrow bounds of the nuclear family. As they described themselves as 'not maternally inclined', they saw advantages in only fulfilling this role part-time.

As a more general strategy, many managers talked of mixing with 'their own kind':

'I've got a group of single freaky friends who no-one would want anyway.' . . . 'I mix with people like myself who are all unreliable and unconventional and don't mind eating meals at midnight.'

This clannishness has benefits and disadvantages. It is easier, more satisfying and a source of social support to spend time with people who share ones experience, interests and values. Typically, however, members of the selective group become more alike and more different from others as a result. The original demarcations and gulfs of understanding and tolerance between them and other social groups are therefore maintained and reinforced.

Some managers' tendency to mix together was an alternative strategy to that of limiting their contacts with other working women, mentioned earlier. It was likely to fit with rather different situational factors, and tended to happen outside work where their behaviour would not be noted and disparaged by male colleagues.

Earlier in this chapter I suggested that one way of reconciling conflicting aspects of identity was to work through an, often painful, new awareness of oneself as a woman in a man's world. The young, single managers' accounts showed that there is an alternative, paradoxical, but no less stressful, route to developing a strong sense of self-worth. Several who had been deterred by the difficulties of developing a viable social identity, had found that by allowing themselves to grow in the opposite, work, direction, they had achieved this goal unexpectedly. Success at work had eventually been the foundation for a base of personal confidence: 'I'm aware that I'm as good as, if not better than, a lot of other people doing my job. That's nice, that's solid. I feel I must be quite competent and that's nice.' This acted as a medallion to take into uncertain social situations, a confident identity which dispelled anxiety about developing other parts of their identity separately and deliberately. Several managers who had initially found the social world daunting had now developed a generaliz ably 'useful' work-based identity — both surprising and delighting themse with this discovery. They were no longer willing, nor needed, to play apologize for or defend who they were and what they did, but sa something solid to hold on to when feeling misunderstood or talked down to socially:

> 'Also, when you're doing a job that is faintly interesting you gain a little confidence because instead of constantly having to talk about little Tommy . . . and be thoroughly bored . . . at long last you can actually talk about something you're doing which is interesting — it helps.'

Another interviewee who is 'no longer alarmed' at going to dinner parties alone says: 'If anyone patronizes and says "Oh what do you do, dear?", I know answering icily will shut them up. You've got something sitting there which means you know that people aren't able to put you down.'

The managers' experiences of problems and successes in relation to social identity reveal paradoxes about trying to be female, or 'normal' as several put it. Many of the idealizations they have striven for are caricatures of femininity — soft, deferring, elegant, relatively incapable — and highly incompatible with their emerging and strengthening work-selves. Their attempts remind me of Cornillon's (1972) warning that, whilst men think women are born with femininity, women know that it is something they need constantly to construct and to strive after. It is ironic, but reassuring, that several managers were developing confident, female identities *despite* their earlier attempts to reconcile themselves to social stereotypes.

V: MANAGERS' IMAGES OF THEMSELVES IN THE FUTURE: BEING OLD MEANS BEING HARD

The managers' dilemmas about identity polarized most clearly as they looked ahead to imagine themselves in the future. They dwelt on who they did *not* want to become.

Many expressed reservations about carrying on being managers 'for ever'. A few felt that the competitive pace would begin to tell as they reached their fifties, and did not relish the idea of trying to keep up with younger competition. It is as if they expected to run out of strategies for being effective at work. Perhaps there is some basis for this fear for those whose management style denied them tactics based in competition.

There was a widely held belief, amongst participants of all ages, that older women managers are 'hard', and a subsequent concern that this might happen to them. Older managers were also expected by definition to be lonely, as they must have neglected their private lives to achieve so much at work. This was another incentive for the sample to try to attend to all sides of their lives:

> 'I am not a career woman. It wouldn't be my choice. I never intended my life to be set out this way. I don't particularly admire career women because it worries me when I feel I might get hard. I think of my family, and that's where I want to be.'

'I do look at older women in our company and try to work out how not to end up like that.'

'It is so easy for the power to go to your head. I have seen it, they are just not the same people. You look at them and think, "that is not the girl I knew".'

Surviving in management was, then, seen as a direct threat to ones identity as a woman. These fears were held most powerfully by managers who already saw themselves changing in directions they regretted. Mainly to cope with work overload or colleagues' competitiveness, they were supplementing their preferred low-key management style with 'male tactics'. One woman disliked the 'toughness' she was now capable of. She had learnt not to be too trusting, was more critical and 'sees through people'. A resulting strain was that she had 'to be careful to have the good manners to listen to them'. Another said, 'I feel bad that I don't have time anymore to be pleasant'.

The managers could look to few women in senior posts to contradict the distasteful image of hard loneliness for them. As the literature on women managers repeatedly points out, they lacked role models to epitomize alternative ways of being. Styles this sample seem disposed to respect and emulate are by nature unlikely to attract attention. Women who adopt an aggressive, masculine style of being effective will be more publicly visible than those who favour a low-key, interpersonal approach. Styles which are compatible with traditional female values are by definition quiet, invisible, likely to work indirectly through other people and unlikely to attract direct attention and praise. Individuals might find personal 'solutions' in ways of being along these lines, but these examples do not speak out to other women. Certainly any woman they could admire was eagerly seized on and described by the sample as an example to be emulated. One manager had recently seen a financial expert on television who crystallized for her how a woman in business should be: 'She knew her subject, was very relaxed and was only intense when it suited the argument'. Some managers also had the double benefits of a woman they admired being ahead of them in their own organization. Not only could they be copied, but they had also pioneered new opportunities and influenced attitudes:

'It sort of rubs off on you . . . I was amazed at how people accepted me from the very beginning. They were prepared to listen and accept new ideas'.

'I think people like her have done a great deal in the company, because she is so level-headed and has helped women's positions very much. Women who have been given senior jobs have done very well and the company realize that women are capable.'

Several managers in the sample were evidently fulfilling similar pace-setting functions in their companies.

It is appropriate here to note how little the managers had to say in about other women at work. The comments they did make show how difficult relationships between women become in the face of the informal organization factors discussed in Chapter 4. Other women were far more likely to be described as difficult to mix with and disapproving of women in senior positions, than as allies or companions. Various circumstances contributed to this. Some women were said to be wary of working for a woman in case she lived up to the female battle-axe stereotype. Others expressed open disapproval of working parents. The managers themselves saw contact with other women as dangerous. The lone woman at a management level risks compromising her special status if she mixes with people lower down the hierarchy. She is relatively safe as the only female — 'they (men) don't see one woman in twenty as real competition'. Even if there were women at their own organizational level, the managers tended to avoid their company for fear of being identified with women as a group, and so obstracized by male colleagues. At the same time several of the sample believed that asking a man to share their lunchtime more than once would be misconstrued as having sexual implications. At this interpersonal level managers therefore felt discouraged from mixing with other women, and from using each other as supports and confidantes, or as a group with potential to influence the organization's culture.

Whatever the richness of their self-concept in the present, interviewees found it difficult to look ahead into the future and trust that working as a manager would allow them a satisfactory sense of self. They expected length of time in such employment to shape them towards stereotypes of masculine behaviour and even further away from images of femininity. As they looked around them in their organizations and saw only the public faces of those women who had been successful in career terms, their worst fears that senior positions necessarily mean being hard and lonely seemed to be confirmed. The estrangement of many from any supportive contact with other women reduces even further their perceptions of possible options.

VI: CONCLUSIONS

Being a manager in organizations as they currently operate puts a woman's identity at risk. She is under pressure to mute her female sense of self and instead to display her abilities to do and act, and be cognitive. As these agentic aspects of her being develop and strengthen, her base in communion recedes. Most of the arenas she engages in distort or deny significant aspects of her identity. At work she will tend to be sanctioned for woman-like behaviour, and outside work few people are likely to recognize or value her professional competence. One strategy for handling her resulting fragmented identity is to mute her awareness of being a woman, to take personal responsibility for patrolling the boundary of acceptable behaviour. In doing so, the manager may achieve her aim of being seen by others as a competent person rather than woman, but she is also likely to find herself cut off from significant aspects of her identity and capabilities. Typically she will have a sense, however, of an energy that is more female, a base of strengths she has because of her gender. Where possible she will draw on these, and express the values of authenticity, cooperation and openness in which they are grounded. Depending on her past experience, personal development and the contexts in which she operates, she will have many or few opportunities to do so. Many of the managers I met reserved their female characteristics to a private sphere. If this expression, too, was denied them, their needs to nurture a core aspect of their being went unmet.

Anchors in close personal relationships — with a partner, old friends or people of one's own kind — help balance and sustain a manager's own sense of self. Conflict and paradox afford alternative, initially more stressful, routes to a whole identity. Some interviewees had been unable to avoid confronting the meaning of their womanhood, and of the difference between them and their dominant environments. Struggling with this awareness, they had arrived at radically new perceptions of the world and of their place within it. These had provided the roots for a new sense of themselves, incorporating an assertive core of femaleness. Other managers seem to have arrived at a sense of their personal worth by paradoxically pushing hard in the opposite direction. They had initially found being a woman socially difficult, something they had to try at. But devoting themselves to their more comfortable work identity had had surprising repercussions. They found that the confidence they subsequently developed was transferable to other settings, and gave them a new liberation to create their own meanings of being female.

These findings prompt me to speculate about current developments. Most of the managers said that they wanted to 'retain their femininity'. I have often heard men express the same wish — of women. I think they may be talking about different qualities. Women are creating new meanings of femaleness which incorporate and value their competence. Men have little if any access to this process, and still draw heavily on stereotyped notions of a constraining social role for images of what being a woman means.

In their dilemmas about identity, the managers have few suitable models of womanhood to refer to. They set themselves apart from women as a general group and have departed significantly enough from traditional patterns for this to be a partly meaningful distinction. Their contact with potentially similar women at work is restricted by the dynamics of their token status. Most interviewees would not identify with the reformist and radical alternatives offered by the Women's Movement either. Too strong an awareness of womens' social position would jeopardize their acceptance of current organizational norms.

During this chapter I have identified the various pressures women experience which make it difficult for them to develop and consistently maintain a sense of their own worth and capabilities. The strength of traditional stereotypes as standards of femaleness and maleness against which they and others judge them haunt the various sections. Ironically women's authentic coping responses, based in communion and acceptance, reinforce and perpetuate organizational aspects which punish them as female, at least in the short-term. But as some move through the obstacles to creativity about identity, starting either from communion or agency as an initial dominant tendency, they develop female ways of being which are new, robust and expressive of women's values and experiences.

This chapter has explored choices which are only partly accessible to the individual's influence. The next, and final, perspective on managers' lives offers more opportunities for deliberate control through balancing different lifestyle options.

Chapter 8

Lifestyles: 'Leading a Balanced Life'

This chapter continues the theme of choice which has underpinned its two predecessors. It focuses on how the managers viewed, and chose between, different lifestyle options. Whilst work was a major source of identity to them, it was but one core life role amongst several. Most interviewees did not see the major options of work, marriage, motherhood and a social life as mutually exclusive, but as mapping a realm of possibilities from which they could select an individualistic combination. Choosing an acceptable lifestyle meant choosing how involved in each of the different roles to be. They particularly identified as issues:

(1) how important work should be relative to other life areas;
(2) how to cope with necessary tasks such as housework;
(3) whether to marry;
(4) whether to have children;
(5) how to preserve some 'free' leisure time amidst a busy life;
(6) how to achieve balance between competing priorities and demands.

The managers' choosing was not usually a conscious process. Older managers' choices, particularly, had mainly happened casually and gradually. They would typically look back to find that, without deliberate intent or awareness at the time, they had moved up the management ladder, had not married or whatever. Younger managers were certainly more aware of the options open to them, and of alternatives to their current positions. A few did plan their futures deliberately, but these were in the minority. Most had some immediate goals, but wanted primarily to preserve long-term flexibility.

All of the managers, but with different degrees of intensity, wanted to lead what they called 'a balanced life', deriving satisfaction from a range of potential roles. I shall look at their lifestyles through an appreciation of what this meant and how they sought to achieve it. My main headings are:

I: Managers say you have to choose

The sample insist that because of the wide range of available lifestyle options, women have, whether consciously or not, to be selective.

II: The various meanings of 'leading a balanced life'

III: Phases of imbalance

Most managers derived their conviction that balance is necessary from times when work had either taken over their whole life or offered too little stimulation.

The next three sections look at the main threats to balance:

IV: Overload: very often being very tired

V: Marriage: achieving balance as a couple

VI: Motherhood

VII: How successful were the managers in leading balanced lives?

Few people achieved their ideal of balance as a routine lifestyle.

VIII: Tactics for achieving balance

IX: Reflections on the balanced life

This chapter brings my account of thirty managers' lives to a close. In a further section I therefore look across the research data as a whole, summarizing the women's approach to employment it reveals, and assessing its contribution to this book's development:

X: Women-centred approaches to employment

I: MANAGERS SAY YOU HAVE TO CHOOSE

In the managers' opinion, women who want to work have to make choices. For some this did mean selecting one role and therefore foregoing others; for most it meant recognizing that they could not engage in *all* their potential roles wholeheartedly and had to compromise. The following comments are typical of many more:

'A lot of women still expect to be able to win in all roles, and I don't think you can, quite honestly.' . . .

'I admire tremendously anybody who runs a job and a family; it's something I've chosen to live with and not to do I suppose. I admire people who do it.' . . .

'I would have done something about it (getting married) if I had been worried. Working women can't complain, the choice is theirs.' . . .

'There are prices to equal opportunities; you can't have it all your own way.' . . .

'Initially women have to work ten times harder than men. I did in the early days whilst my friends were going out enjoying themselves. But everyone has to if they want to get on. I wasn't the gay young thing. You have to make sacrifices. I was too tired, and I still am.' . . .

'If you want to succeed you have to be prepared to sacrifice quite a few of the normal things, but then all life is trade-offs, isn't it?'

Most, then, did not think they could 'have it all' (Pogrebin, 1978). Marriage and social life are the main areas of potential sacrifice if they chose to devote substantial energy to work.

Generational differences in their answers show two trends. One is towards more deliberate choosing. Older managers were more likely to talk of making the most of paths they had unreflectingly followed to the exclusion of others, younger managers of appraising the options more consciously as they went along. Secondly, some older managers saw marriage and work as mutually exclusive. Younger colleagues said they wanted both to some degree, but are highly aware of the potential obstacles to doing so.

Whilst combining marriage and work was increasingly likely to be seen as practicable, having children as well was considered difficult. Three of the sample had recently had children, and others hoped to do so later and to continue working. All had been, or will be, relatively old first-time parents. They are part of a generation of workers who have decided to have children once they are relatively established in a career. This was not a plan with which they started their working lives. Rather, their greater consciousness of the options open and closing to them prompted many to consider the issue before it became too late biologically. The mothers in the sample had therefore decided to have children at relatively short notice. Other managers were doing more advanced planning, as far as they felt able in this highly uncertain area.

Younger managers, whilst agreeing with older colleagues that some choosing is necessary, conceived of choice differently. They were moving away from seeing options as probably mutually exclusive. Their notion of choice was grounded in a concern for balance. They did not want to maximize all possible lines of development, but neither were they prepared to emphasize one life role to the total exclusion of others.

II: THE VARIOUS MEANINGS OF 'LEADING A BALANCE LIFE'

The balanced life to which the sample attributed so much importance had a multitude of connotations. For the manager herself it meant not devoting all her life to work, preserving a satisfactory private and social life, having time to relax, avoiding stress at work and physical strain generally, and not neglecting responsibilities she had in other important roles. Typical expressions of these values were:

> 'I don't want to be a person who does nothing but work. I've got too many interests, obligations and satisfactions outside.' . . . 'I want to be satisfied in three thirds of my life.'

The notion also stood for retaining her sense of personal identity, including its foundations in being female, in the face of a lifestyle which, in others' views, and often in her own, contradicted and threatened it. Chapter 6 showed how these concerns contributed to decision-making about work. In terms of their lifestyle overall, the managers assigned identity a high priority, which was reflected in the wide range of potential options they considered. Despite their involvement in work, for example, several were prepared not to work at some time, or to move into independent or lower status jobs to achieve a more satisfactory sense of self. Jobs which threatened their sense of themselves as women would be rejected, or at least not openly sought.

In their dealings with others, some managers also demonstrated feelings of responsibility for promoting balance in social situations and life generally. At work, as we have seen, this meant fostering a gentler atmosphere, paying attention to others' needs and countering rather than contributing to competition and aggression. Outside work most of those who were married felt personally responsible for keeping the family's home running smoothly, and for managing the various sources of potential overload.

One common thread ran through individuals' varied interpretations of what leading a balanced life means. That thread was having time, particularly for the nurturing activities in life such as being pleasant, spending time with people, making settings pleasant, and having a social life. The managers valued these activities because they formed parts of a cluster of deeply rooted images of who they wanted to be. All too often the pressures and pace of the lifestyles they lead made taking time over anything seem impossible.

In the next section I report the forces for and against balance managers experienced, and learn more about the managers' ideas of what a balanced life looks like and stands for.

III: PHASES OF IMBALANCE

Most managers referred to critical phases in their lives which had led them to new understandings and resolve about what lifestyle they wanted. They gave examples of moves both away from and towards greater work commitment, but the former were far more frequent. About a third of the sample described particularly stressful times when their jobs had taken over their lives. They reported the strain, tiredness and disrupted relationships this had caused. The following are typical examples:

'I used to worry terribly, but now I've learnt to close the door behind me. I got to a desperate stage, but I realized I must either be sensible or get out. Seeing people die early played a part. Nothing can be that important or serious.'

'The job used to dominate my life, be a major interest. At one stage I felt that was all I was doing. I didn't feel that was quite right, I was getting very introverted. Now I've got over that.'

'When I first took this job on my whole life was dedicated to work. I thought here's the opportunity . . . I've got to put everything aside until I can do it properly. Having done that I got into the routine and then I couldn't get back . . . I couldn't understand why it was that having been a golf addict suddenly I felt, "I can't be bothered, I'll go tomorrow". Eventually I analysed it and realized that, with the pressure of work, when weekends came round all I wanted to do was collapse.'

Their accounts of going through cycles in which initial challenge and experiences of stress lead eventually to mastery and competence are similar to those reported by other people I have talked to who are moving up organizations and taking on increasingly demanding jobs with each promotion (Marshall and Cooper, 1979). The interviewees' reactions, particularly their awareness of potential health costs and their resolution not to repeat such episodes, are less common. As the examples show, they were pleased that they had achieved a detached view and resisted a potentially stressful lifestyle becoming too firmly established. Most reported that they had subsequently arrived at 'a new perspective', and that work was less important, relatively, as a result. Self protection seems a key theme here. It is interesting that managers did not assume they would be able to cope more competently in the future as a result of their experiences. In relation to work their learning may initially seem unnecessarily short-term and non-transferable. But they were, again, taking a different, broader, perspective. They derived significant lessons for their lifestyle as a whole from these stressful interludes.

Other critical incidents had moved individuals towards greater work commitment. Some managers had experienced short periods out of work, for example, which had served to confirm, possibly even reveal, its true importance to them. As one put it, after one such break 'I decided to get back into being ambitious again'. Another fairly typical account from a manager who had taken maternity leave illustrates how this unintentional trial showed both her and her partner the sense of their more usual lifestyle:

182

'I found it very difficult to adjust to being home all day. We found we got into the classic situation: I hadn't talked to anyone all day and was bouncing around wanting to know what he had been up to. He would stagger in exhausted and say "shut up". Neither of us liked that . . . He disliked the idea of me being at home as much, if not more, than I did. Because that just wasn't how he knew me.'

Other people had found that, as they had been moved up their organization, performing more responsible jobs had changed their initially neutral attitudes to work. Theirs was a more gradual realization or realignment. One manager who had initially adopted a relatively casual attitude said that as she was still working rather than following a more domestic route, she might as well work a little harder, and earn a little more money.

The notion of a balanced life is so powerful because it is both the expression of an ideal and an identification of the essential threat in the managers' lifestyles. The next three sections explore the main factors which continually jeopardize balance in their lives.

IV: OVERLOAD: VERY OFTEN BEING VERY TIRED

If the managers I interviewed had one concern above all others, it was that of overload, having too much to do in too little time. This was not attached specifically to work, but pervaded their lives generally. As a result they were 'very often very tired', and this had significant impacts on their lifestyles. Several factors contributed to overload.

The managers saw their jobs as highly demanding. The comment 'publishing just doesn't stop', and 'retailing can easily take over your whole life' were typical expressions of the relentless pace and potentially engulfing nature they ascribed to work. Most worked long hours. Most in publishing mixed socially in a semi-business way with colleagues outside official working hours. Business travel was an added pressure in this already busy picture.

Some managers identified the way they did their jobs as a further element contributing to overload. Many, more perhaps in retailing than in publishing, commented on the high levels of personal involvement and commitment they showed. Their work styles relied heavily on submerging their whole selves in a task, as a basis from which to perform well. The grounding they thus achieved contributed to their confidence in doing the job well, and to their ultimate satisfaction. The managers recognized however, that the amount of energy such strategies required probably contributed to the pressures and frequent exhaustion they experienced. One compared herself to her husband in this respect. He seemed to be able to work as effectively without getting as tired as she did. She did not want to evaluate either style as better, but noted their differences.

These comments about work style link to several themes developed in earlier chapters. They provide evidence of a female cognitive style, whose preferred ways of working require more personal energy than alternatives might. They may also relate to the risks women employees represent. One way women reduce these risks is by being very good, rather than average. Doing the job to the best of their ability earns them the acceptance men can achieve through only satisfactory performance. From watching women throughout their organizations and reflecting over their own experiences, the sample endorsed the old adage that women have to work (two/ten/several times) harder than men to get on.

Several single women in retailing felt they could not marry and continue working as they did because of their style's demanding character:

'I haven't yet reached the stage where I can settle down, because to get the job done as I do I have a certain involvement. I can't imagine being married until I have reached a level where I am satisfied and have reached the stage I want to get to and can cope with it. It wouldn't be fair to anyone.'

'Because I get the job done as I do, I have to have a certain involvement.'

'I do not feel there is any way I could do this job if I were married because there would be a conflict of priorities. I wouldn't be able to do it to my satisfaction, let alone the company's.'

The managers were, as we have found before, unsympathetic to changing their style, despite its costs. Although some of the environmental pressures, such as prejudice against them as women, which might have helped foster it, had moderated, these were not the only considerations. Most interviewees rejected ideas of doing their jobs with less expenditure of personal energy because they would risk becoming superficial and unauthentic. These designations were incompatible with strongly held values about how they would like to behave.

Individuals' circumstances outside work varied. I shall explore some of the major differences below, but there were also many common features. They all wanted to maintain other aspects of their lives such as social, leisure, sports and cultural activities, but found doing so difficult. Most experienced housework as an additional drain on their time and energy, and therefore a pressure. This was a concern for both married and single managers, but was a more pressing problem for the former. Housework became particularly burdensome to the sample in relation to special occasions. Some, for example, were in jobs which appeared to call for entertaining at home:

'I ought to give a lot of dinner parties but I give none. I don't know when I'm supposed to do the shopping or cook it, I've no idea. In theory I could employ someone to do it, but I think people think it is less hospitable of a woman to have a meal cooked, and less friendly — as though you aren't making an effort. Whereas, if a man's wife cooked it, that's something the man's wife does.'

Repercussions of the general life overload the managers experienced, and evidence of its severity, can be seen in how they spent their time outside work, and the kind of social lives they developed. Most did very little during the week apart from work, unwind, cook an evening meal and go early to bed. Weekends were hardly more hectic. They might see friends and relatives. Most were selective socially, preferring to spend their scarce time with people who shared their values and priorities. Doing housework to catch up on the previous week, and prepare for the next, was also a common activity. The managers' answers when asked directly what their 'satisfactions outside work' were, revealed that they had very few other spare-time activities.

Marriage and/or parenthood add additional roles to the managers' potential repertoire. They therefore increase the potential for overload, especially by multiplying the number of intersections between roles at which conflicts can occur. The following sections explore the sample's experiences of these relationships.

V: MARRIAGE: ACHIEVING BALANCE AS A COUPLE

Half the sample were married or living with long-term partners. For convenience I shall use the terminology of 'marriage' to designate all these stable long-term relationships. This group had more stability in identity terms than their single colleagues (see Chapter 7), but had to handle a more complex array of demands and responsibilities. The managers' discussions of their partnerships centred on how they interpreted their role of 'wife' and how they achieved satisfaction of their own needs as part of a couple. Two key topics illustrate the significant factors in operation: their partners' reported attitudes to the managers' work, and the division of responsibility for housework.

Partner's attitudes

How their partners saw the managers' work was very important to them. They wanted acceptance of, and hoped for support with, what they were doing. Partners' reported attitudes to the managers' work varied along a spectrum from actively approving to reluctantly tolerant. More were towards the positive end of the scale. This was welcomed by the managers, many of whom said that they would find it difficult to carry on working unless their husbands did approve.

Most managers said their marriages were founded on their identities as people who work, and that combining these roles was natural for both them and their partners:

> 'That's the way we met, the way we know each other, the kind of marriage we have.'

> 'I'm not lucky in my husband or whatever, but he sees it as utterly natural and desirable for me to work, and he's rather more keen that I should go on working when I have children than I am.'

'I don't think my husband can conceive of me not having a job . . . We met each other through work, we combine very much on a work–intellectual–business level, as well as a personal level. I think both of us would be really very worried about what it would do to our relationship if I stopped working.'

For some couples the acceptability of this pattern of relationship had been tested when the manager had a brief period out of work. This usually unintentional trial had confirmed both partners' preference for their current arrangement.

The multiple strengths of such relationships, which had as firm an intellectual as an emotional base, were emphasized. Neither partner relies on the other for worldly stimulation, each can understand the other's interests, triumphs and problems, and they can encourage each other or provide a balanced perspective as appropriate. Several single managers certainly identified these as benefits they lacked.

In their descriptions of partnership, these managers were voicing strong preferences for symmetrical relationships. They had departed significantly from traditional models of marriage in which wife and husband complement each other in characteristics and roles. Their new pattern had implications for their performance at work. Several managers (mostly in retailing) portrayed their partners as home-based mentors. In this role they had given the managers' careers significant boosts, by being more confident and ambitious for them than they had been for themselves:

'He's very anxious for me to go as far as I can. In fact, I think he has more faith in me than I have in myself. If it hadn't been for him I might still be at a dead-end in my previous company.'

Several managers also noted warmly that their husbands were 'proud' of their achievements; others that their partners were more stimulated by the interviewee's job than they were by their own. There was some evidence in a few managers' accounts that they and their husbands spurred each other on to greater work involvement. None, however, appeared to have the stress-provoking preoccupation with bettering each others' achievements predicted in some of the literature about dual career families.

Other husbands were less keen for their partners to work — their attitudes could best be described as 'tolerant'. In these circumstances managers were unlikely to discuss work at home, and tended to lead more fragmented lives. Usually they felt under pressure to take less interest in work and spend less time on it than they might have liked. Taken together, these managers' own dilemmas about leading a balanced life and their partners' disapproval of their work left them uncertain about their own needs and wishes. In retailing, some less enthusiastic husbands were said to appreciate the household's second income and the luxuries this brought. The same certainly was not reported by people in publishing, who repeatedly told me how low pay is.

Most managers wanted their partners to approve of them being employed and of the kind of work they did. Those who had continually to defend their desire to stay in demanding employment found this a great strain. To resolve and anticipate difficulties in their lifestyle, many managers took on responsibility for maintaining the couple's balance. They would be watchfully attentive to their relationship and to signs of potential stress and overload. A few said they had at some time anticipated conflicts between their own and their husband's careers and had moderated their own development to avoid this. Typically this moderation was expressed through inaction rather than action on their part. They might coast along at a particular organizational level, for example. Only very occasionally did the organization interfere with this behaviour by offering them promotion. The managers explained their behaviour in terms of the practical aspects of overload. They thought that the marriage would simply be unable to cope with all the demands on its resources. For most managers, this had led to a temporary rather than permanent accommodation. Once they felt circumstances were more favourable, they allowed themselves to become more involved in work and enthusiastic about further development opportunities.

My discussions with the sample about housework went into more practical detail, and provided further evidence that most felt personally responsible for guarding the family's balance. Some managers appreciated, however, that in doing so they were acting from perceptions of their role and duties which they thought outdated and which their partners did not wholly share.

Housework

One of the more practical matters we discussed during interviews was who does the housework in a dual career marriage. The available literature paints a relatively bleak picture in this area from a woman's viewpoint. From their extensive survey of research on who does what in homes where both partners are in employment, Gowler and Legge (1982) conclude that women take the strain with some *help* from their husbands. The data explain Langrish's (1981) research findings that women workers saw home life as a pressure. To them it represented hard work and little time to themselves. Men, in marked contrast, saw home as a refuge, a place to recharge their energies.

The managers in my sample were dependent on their partners' attitudes for how much housework was shared. They reported a broad range of participation, from complete to minimal. Most interviewees described housework as a joint venture and felt fortunate in this sense. Many further reduced potential demands by relaxing the standards they considered acceptable. Some difficulties did still occur, however, mainly because the family system as a whole was grossly overloaded.

Given this discrepancy between what they hoped to achieve and what seemed practicable, managers again stepped in themselves to bridge the gap. Most married members of the sample said that they felt responsible for running the house, and guilty when things failed to go smoothly. These feelings of conflict made them reluctant to ask assertively for help. Whilst recognizing the inappropriateness of these feelings in the symmetrical marriages they applauded, many were unable to escape the powerful expectations they had of themselves:

'I feel I ought to be superwoman and do everything, but shouldn't have to apologize if it isn't done . . . it's a strain at the moment.'

'He does just as much housework as I do. He doesn't expect me to produce three-course meals promptly on the dot . . . if I don't manage to do the shopping we just have spaghetti and that's it . . . Although I've said, and its true, that we do share the housework, I do feel a pressure which I can't resist that it ought to be me who does most of the housework and the cooking. I do feel guilty if the house is looking like a tip and we've had spaghetti three evenings running, and I find it an amazing strain to find the time to do the ironing, which is the one thing my husband won't do. You could say ironing was the greatest strain on my life.'

Only in one case did the basic assumptions that housekeeping is women's work seem to have been totally reversed. One manager's husband was less involved with work than she, and very active in home and private life maintenance. She said, 'he balances life out for me'. A telling phrase.

VI: MOTHERHOOD

Only four interviewees had children. One's family were in their mid-teens, and the rest were small children. Several others hoped to have children at some stage and had considered the issues involved. A few managers in the sample said they might have had children had things turned out differently. For most this meant if they had not worked continuously in jobs they found interesting and stimulating. Only one expressed significant regrets that she had a career rather than a family. Only one working parent came from retailing. The sample is far too small to call this a significant industry difference, but other indications do suggest that managers in retailing were less likely to want, or to be able to find time to have children.

For those considering the matter deliberately, whether to have children at all was usually open to debate. But this was one decision on which the managers did not feel able to arrive at anything like a rational conclusion. 'It's a major decision but you have nothing to go on . . . in the end we just took the plunge', is a typical comment from one of those who had recently become a parent. The recent parents I spoke to had not really considered the option of giving up work. They had always assumed they would continue, and felt this was wholly appropriate to their previous experience and who they were as people.

Once the decision had been made, timing was a further problem — 'there's never a good time, and always reasons to delay'. One issue for managers who wanted eventually to have children was whether to inform their superiors or companies during discussions about job prospects. Most felt they should keep their plans strictly to themselves. They saw these as 'private' details, and thought their career prospects would be seriously damaged if they suddenly became a special class of employee in this way:

> 'I don't want to tell anybody about it because if it doesn't happen I still want my career . . . I learnt very early on that you keep personal life very private and you don't tell anybody about it, because you want them to accept you on the same basis as everybody else.'

Whilst maternity leave is now readily available, jobs at the senior organization levels the managers worked at were difficult to leave and come back to. Two mothers had been luck enough to find ideal temporary substitutes — people who were about to retire or leave the company — and had been able to keep in touch with work relatively easily. Another found taking maternity leave practically impossible, fearing she would never catch up again. She had virtually worked through, 'but slowly', and had found this physically and emotionally exhausting.

Comparatively speaking, I was told, the 'real' problems started after maternity leave. All parents in the sample were employing nannies to look after their children whilst small. A couple of the aspiring parents in retailing would prefer to stay at home with young children to have an influence, but this would only be a short-term measure. On the whole, employing a nanny seemed to work well and the emotional complications some had anticipated of being jealous of the child's affection for the nanny had not materialized. The crises came when nannies left — which they did quite frequently. 'Nanny', it seems, is very much a passing trade.

In relation to childcare, too, the managers took on responsibility for coping with any difficulties caused by their lifestyle. Some reported occasional feelings of guilt that they should be at home with their children. These conflicts presented an additional pressure for them to manage:

> 'I think its still a conflict of me feeling sometimes I ought to be at home. Although if I look at the situation properly, I don't want to. Its a conflict I have to cope with really'.

All asserted forcefully that they would give up work should the need arise, but hoped that it would not. The family came first, their jobs were important but secondary. This typical comment shows the conflicting forces in operation:

> 'Obviously in the final analysis, if there were real problems, I suppose I should have to stop. He is our responsibility . . . It would be irresponsible of

us and unfair to him if I determinedly ploughed on with the job. I couldn't do it if I didn't feel completely confident that he was fine.'

Multiple role as threats to the balanced life

The literature on employed women interprets many of their difficulties as the result of conflicts between multiple roles. This sample's accounts of overload in their lives show that the nature of these roles was at least as significant as their number. There are several contributory factors. The first is the managers' guiding aspiration. By seeking a balanced life, they defined several life roles as approximately equally salient. They could therefore not cope with conflicts by the relatively effortless strategy of giving one role precedence over others. Instead they had continually to balance priorities. Secondly, the managers tended to take responsibility on themselves for achieving balance for their marriages as well as for themselves individually. When their lifestyle or that of their partner or family came under pressure, they felt they should manage the disturbance. Even if other people signalled that this was unnecessary, they were often unable to escape feelings of guilt that they personally should be doing more.

A third factor contributing to the build-up of overload in managers' lives, was the high standards of performance they set themselves in most of their main roles. Whilst some compromised about housework, no other life areas offered similar flexibility. Finally, managers' lives were made more vulnerable to potential stress by their tendency to define themselves in relation to other people, rather than individualistically, and in ways which left them responsive to others' demands and evaluations. Many of their positions, officially or unofficially, were 'service' roles. Adopting this approach exposes the manager to drains on her time and energy as she will often be dependent on others' availability, needs and cooperation. Roles thus defined also leave their holders open to frustration. Their objectives are too vague, indirect and long-term for achievements to be identified easily, and the individual's efforts seldom attract attention or recognition.

The roles the managers created outside work express tendencies towards communion particularly clearly. Not surprisingly, the managers' preference for interdependence was even stronger in this, their traditional, domain than in employment. Again we see how vulnerable it left them. At home they seemed to be less able than at work to protect themselves by adopting 'male tactics'. In their intimate relationships they wanted, and were expected, to reveal their female traits more clearly.

The various threats to balance in their lives stretched managers' abilities to cope. Often-conflicting undercurrents of stereotypes, self-perceptions and values could not be easily or permanently reconciled, and managers were continually faced with dilemmas and choices. The next section shows how these affected the dynamic shape of the managers' lives.

VII: HOW SUCCESSFUL WERE THE MANAGERS IN LEADING BALANCED LIVES?

Given the challenges they faced, it comes as little surprise to find that the managers as a group were only partially successful in their attempts to lead balanced lives. They were 'very often very tired', but many had achieved a new equilibrium in their lives after a previous phase of stress. Only a very few (older) managers appeared to have achieved balance as a steady state. For the rest *balancing* was a continual, vital process in which they were engaged. Their accounts suggested that many managers' lives were a recurrent cycle of moving into imbalance and back to balance again.

The managers themselves tended first to notice movement into this cycle as their lifestyle drifted into imbalance. Most often this meant becoming more involved in work than previously. They would suddenly recognize that this had happened by the additional strains they and/or their marriage were experiencing. Prompted to analyse their situation, they re-evaluated their priorities and how these were reflected in their behaviour. The manager then acted to redress the balance, by deciding not to bring work home, for example, or experimenting with a new style of working, and achieved a new equilibrium. Typically this new state would also be temporary. In time drift would occur again, and the cycle restarted.

A manager in retailing, for whom Saturday working was now optional, described the insidious moves into imbalance as she typically experienced them:

'You get to the stage when even if you are home on Saturday you feel guilty. That is the sort of pressure that builds up inside you until you are only really happy when you are at work. You have to try to reverse that, otherwise its a problem to unwind and life becomes difficult at home.'

One, at least, of the sample was at the unbalanced stage and pausing to reflect when I met her. Her comments indicate how fundamental the lifestyle review will typically be:

'Somehow I have to bring things back to a better level than they are now . . . I have even thought of stopping work altogether or doing something completely undemanding. I don't know whether it would help; I think I would just get bad tempered.'

The balancing cycle's phases brought their own intrinsic pressures. In addition, failing to match ideals of the balanced life was a source of dissatisfaction and disappointment in its own right. This was a core standard by which the managers judged themselves. Within this framework, individuals drew on a range of strategies to sustain or re-attain a satisfactory lifestyle. This chapter's earlier sections have considered the major options open to them of whether to marry and whether to have children. The next looks at the tactics they used for achieving everyday adjustments in lifestyle.

VIII: TACTICS FOR ACHIEVING BALANCE

The sample drew on an array of life-management tactics to foster the balanced lives they sought. Which tactics they chose and how they used them were determined by their individual interpretations of what a balanced life meant. Two dimensions of variation run through the range of tactics listed below.

For some managers balance meant leading an integrated life in which work and home merged into and influenced each other; for others it meant separating these life areas off from each other. Both strategies offered their own inherent advantages and potential pressures. Which shape in these terms their life tended towards was only partly a matter of the manager's personal preference and control. A web of other influences also played its part. The few people in publishing whose partners also worked in the industry were the most likely group to lead integrated lives. They talked freely about work at home, and mixed mainly with other people in publishing. They found their lives caught up by and rooted in the industry. This had many advantages, particularly their knowledge about and involvement in what was going on and their acceptance into the industry's informal social network. It did, however, mean that they sometimes felt that publishing was taking over their whole lives, and that they could not escape to see the world from other perspectives. At the other extreme were managers who completely separated work and home. Circumstances which encouraged this were working at a retail store which provided few potential friends, developing separate and incompatible work and home identities, and having a husband who was only 'tolerant' in his attitudes to the manager's job and refused to mix with her work friends. Being able to relax, unwind and develop other interests in a separate world were seen as benefits of this pattern. Having sometimes to mute their work identity, and having more competently to manage conflicts at its boundaries with home, were major disadvantages.

A second consideration was how much flexibility the manager's lifestyle incorporated. Adopting tactics which preserved flexibility made it easier to cope with unpredictable and changing circumstances. The alternative approach of organizing and controlling helped to fix in place the particular manager's personal interpretation of balance. Again a broad range of influences were at work. At a particular lifestage some life areas became almost solid, offering no room for adjustment. For example, parents felt that they had to be particularly careful about organizing their time effectively. Partners' expectations, having children, and their own high standards of performance were the three main factors which reduced the managers' room for manoeuvre in some or all areas of their lives.

The following tactics are mainly managers' own ways of coping with potential lifestyle pressures. Some who were married reported approaches their partners shared. The strategies range across the two dimensions of integration–separation and flexibility–organization, some promoting one pole of a dimension, others another.

Relaxing standards and expectations

Some managers relaxed the standards they set in certain areas to counter potential overload. This was particularly common for housework. A typical example was the senior manager who tried when first married to be a good spouse and housekeeper, but not to let her job slip as a result. She felt that in trying to prove herself in these two new roles she attempted too much, and later adopted a much more casual approach to housework. She was, however, only able to relax in her home roles once she had proved her capabilities. A similar strategy was 'learning not to mind' about a variety of things. One person, for example, had learnt not to get upset when home plans get disrupted because of work demands. She now takes cancelling holidays in her stride, she said.

Buying-in services

Some managers paid for housework or childcare services to supplement their own resources. From their accounts there seem to be plenty of opportunities for the service industry to expand the range and quality of its offerings.

Selective social lives; mixing with their own kind

Lifestyle pressures reduced the time and interest managers had for social activities. Most spent these sparse resources sparingly, and mixed only with people they liked and who at least respected, if not shared, their basic values. Mixing selectively with people like themselves — single career women, married working women with children, or whatever — provided additional reinforcement and support for their values and lifestyle. Some managers referred to the difficulties, disapproval and sometime open criticism they had experienced outside these selective circles. Those who felt they were accepted and belonged socially were particularly reluctant to move for the sake of promotion. They would perhaps be more mobile if similar roots could be easily established in new locations.

Regulating the boundaries between different life areas

The managers who wanted to maintain segregated lives adopted a variety of tactics to achieve this. Some had a policy of not taking work home; others used a long commuting journey to shift from one world into the other. In a similar vein some managers specifically allocated time to non-work activities to prevent the job encroaching beyond its place: 'We make it a rule that we have weekends off. If we didn't, I don't think we would cope.'

Phasing

Several managers spoke of managing competing demands in the medium-term by temporarily concentrating on one life area as activities there peaked. They would then redress the imbalance later. One had been happy, for example, to mark time in her job whilst newly married. Another was paying renewed attention to her career as her social life had settled into place after a move. Other members of the sample expected to use this tactic in the future, especially to accommodate having children.

IX: REFLECTIONS ON THE BALANCED LIFE

The managers' concern about leading a balanced life was an expression both of an ideal and of the inherent threat in their lifestyle. Whatever meaning they attached to this dictum, working brought with it both pressures to be handled competently, and opportunities to develop aspects of their identity which would remain largely unexpressed through women's traditional social roles.

The notions of balance to which managers were committed are grounded in communion, and show that this modality has a strongly personal reference as well as the altruistic concern about society I have previously emphasized. These meanings are essentially compatible. In their attempts to achieve balance the managers were nurturing themselves (especially their health and sense of satisfaction) as well as the social groups they belonged to. Unless they do so their capabilities as parts and promoters of the whole (family, work group or community) are weakened, and the whole eventually suffers. This material provides tentative empirical support for Chapter 4's suggestion (in relation to power sources) that, if interdependence becomes self-sacrifice and loses its self-sustaining dimension, it is eventually dysfunctional and destructive to all concerned. In the recent history of sex roles women's places in society have indeed become stripped of social and intrinsic worth in this way, and women are losing faith in an underlying complementary order. The low social worth of their contribution has robbed them of personal worth, and they are seeking both in their demands for women's rights.

X: WOMEN-CENTRED APPROACHES TO EMPLOYMENT

It is now time to build on Chapter 6's concluding remarks and assemble the main themes in the female paradigm of employment to which the research material lead me. It balances women's attraction to, knowledge of, and participation in a world of employment shaped by men, with the base of women's values from which this participation is continually informed. I have repeatedly found, in earlier theoretical chapters as well as those recounting individuals' experiences, that these two perspectives coexist and intertwine.

In this synthesis of the three preceeding chapters' material, I shall collect together the recurrent themes which resonate women's values as they find specific expression and development in encounter with employment. This is a provisional step in a continuing process of spinning women-centred notions of employment. The emerging frameworks reveal the challenging opportunities which face working women, their dilemmas, and their opportunities to reclaim their own strengths. The themes below are expressed as ideals. The research material above has shown how difficult many were to achieve in practice, and consistently.

Personalistic choosing

The first recurrent theme in a female paradigm of employment is that of choice, and its nature and shapes for women. The managers chose from a wide range of lifestyle options and set few constraints in making particular decisions. At a conscious level at least, many had a surprisingly low attachment to their current paths of development. Most expressed a continual openness to radical change. The managers so commitedly espoused flexibility because they took themselves, rather than the world of work, as their reference point for co-ordinating choice. They did not start from employment or the role of manager as a given in their lives, but moved from a sense of their own identity towards these as possible options. Their choosing took the form of assembling a repertoire of roles suitable, at a given time, to make up a whole balanced life which matched their needs and values. They were thus continually creating and re-evaluating its shape.

The managers' choices were guided (but not wholly determined) by their strong needs for self-fulfilment, personal development and to be stretched fully by stimulating and challenging work. These criteria are highly personalistic and immediate, and pay little heed to public norms of social worth and status. They suggest new meanings for 'career' and 'ambition' as they relate to women's participation in employment. The managers main 'ambition' was to ensure future opportunities for personal satisfaction and growth through work. Similarly, a 'career' was a sequence of jobs each meaningful and appropriate within an individual's whole life pattern at a particular time.

Their style of choosing gave the managers a sense of personal responsibility, of being in charge of their lives. They used this discretion to decide how committed to be to different life areas. Some set limits to involvement at work or pursuing promotion; others relaxed their standards of housekeeping. For many, feeling responsible also translated into the pressure of trying to prevent other family members suffering negative consequences because they had chosen to work.

Self-expression through multi-stranded lives; the need for balance

The managers, then, sought to derive satisfaction and growth by engaging in a variety of simultaneous life roles. They were not prepared to maximize one to the exclusion of others. Through these various elements they expressed and developed a rounded sense of their own identity in both its communal and agentic aspects. In doing so they valued authenticity, and would reject options, such as particular styles of management, incompatible with their sense of self. This diversity of roles brought with it the need for a limiting principle to reduce potential overload. Managers found this, as well as a key expression of their values, in their ideal of the balanced life.

Contextual awareness; accepting connectedness

Managers twinned their personalistic decision-making base with an awareness and concern about the wider setting in which they operated. They looked to their organizational and social contexts in various ways.

Many had some awareness of being in a male world and being different. Working within a male culture presented them with opportunities and barriers, satisfactions and pressures, all of which they had somehow to manage. Outside work they found themselves different again, but from traditional expectations of female behaviour.

Most managers in the research sample favoured harmonious interdependence in their relationships, and promoted cooperation whenever possible. Those who had reached senior organization levels regretted the more competitive atmosphere that dominated. Even those who prized their independence would prefer to encounter more cooperative relationships than they did. They were disappointed that male colleagues at work misunderstood or patronized them, and that social contacts were often aggressive or derisory.

Particularly in their relationships outside work, managers accepted their connectedness with others, and were willing to experience the duties as well as the benefits this brought. At work, too, they took the responsibility for nurturing cooperative relationships on themselves. By staying with one company many managers also acknowledged the benefits of stable relationships.

Engaging through openness

Openness, my fourth theme, has several elements. It underpinned interviewees' management style. They wanted to establish authentic relationships as a basis for achieving work tasks. Openness also provides a new interpretation of planning, one which is expressive of communal acceptance rather than agentic control. Whilst managers' focus was typically on current job satisfaction, an expectation of change was seldom far away. They wanted to be changed personally by the work they did, and also expected the environment to change around them. The managers' main form of planning welcomed these prospects of change, and was orientation-shaping rather than goal-directed. Managers anticipated possibilities, and often chose options which would help them adapt in the future.

The managers' preference for working through immersion is a translation of openness into a characteristic cognitive style. This approach, whilst demanding in terms of energy, provided the necessary grounding for them to enjoy work, and to perform well and attract attention.

Assuming and perceiving wholeness

In various ways an assumption of wholeness dominated the managers' lives. Their belief in authenticity and honesty represents a particularly significant translation of wholeness into their perceptions of employment. By bringing their private, personal approach with them into organizations, ignoring distinctions between the formal and informal organization, and favouring 'bluntness' which refuses to disguise emotional statements in rational language, women managers are reconnecting dimensions of experience which have become artificially separated by agentic work strategies. These developments parallel the assertion that *the personal is political*, and suggest that women can offer a means of returning to wholeness as one of their contributions to organizational life.

Female values

Underpinning the managers' accounts were values which acted as meta-perspectives to guide the many life choices they faced. These, too, have strong links to the principle of communion, and epitomize the female perspective women managers bring to employment. The most significant values they applied were adaptability, authenticity, balance, co-operation, engagement, interdependence, health, openness, and wholeness. The managers combined an intensely personal view with an attunement to the wider context as their framework for evaluations. In doing so they seldom referred to an intermediate level of external social norms such as established career paths to guide their choosing. Instead they made individual decisions anew each time to draw together the personal and contextual perspectives.

The managers' values in context

The choices the managers made from their perspective illustrate an interesting pattern in the threat they pose to men's position in society. Ironically the decisions women managers seem to be reaching about lifestyle (the aspect of their recent development to which attention is usually drawn) are much less radical than those concerning their identity. The principles of balance, health, nurturing, cooperation and environmental attunement moderate women's potential intrusions into men's public world, and ensure the continuance of caring and servicing functions in society. They are compatible with other people's apparent needs. Women's continuing affirmation of these values does not, however, mean that the Women's Movement has died, but rather that it has found more female-compatible ways to have its effects.

Women's growth and development in terms of identity are more radical, and more likely to disturb the pattern of female–male relationships. Women are questioning, and often rejecting, stereotypes of what it means to be female. There is a major contrast between relatively minor changes in observable lifestyle, and radical shifts in consciousness and personal identity. Whilst appearances stay much the same, the underlying pattern seems to be changing.

XI: TOWARDS IMPLICATIONS FOR WOMEN MANAGERS

Having reached an understanding and valuing of women in their own right, both theoretically and in the more confusing actuality of their lives as managers, I am now ready to draw implications from what I have learnt. As I have repeatedly found, there are two broad alternatives open to me. I can set my sights on reform, accept the current male world of organizations and suggest women survive within it. Or I can question basic assumptions, look to alternative, women-centred value sources, and spin new meanings and patterns of employment. The next three chapters explore different avenues forward. Chapter 9 summarizes strategies commonly identified as ways to increase opportunities for women to take on and progress within management jobs. In Chapter 10, I suggest that these standard solutions are not only not enough, but often do not work as intended. The book's final chapter explores alternative views of potential futures.

Chapter 9

Reform: How to increase the number of Women Managers

This chapter brings together commonly suggested strategies for increasing the number and prospects of women in management jobs. These proposals, many of which are already being tried out, accept organizations as they are, and that men are dominant and provide the model for being managers. They advise women to accommodate, and fit themselves into existing systems and norms, preserving whatever flexibility they can. This chapter is, then, about surviving in a man's world.

I have grouped initiatives and potential initiatives under six broad headings. These are shown in Table 9.1. Some provide formulae for translating accepted interpretations of equal opportunities into practice. Others advocate flexibility and tailoring initiatives to a particular organization's or individual's needs. The first four categories of proposals require change from the organization and its current members; the remaining two consider what women can do to help themselves. I shall now consider the six areas of advice in turn.

I: INTEGRATING WOMEN INTO CURRENT ORGANIZATION SYSTEMS AS EQUALS

A substantial chunk of advice is directed at what organizations in the shape of policymakers and company personnel staff can do to ensure equality during recruitment, training and selection for promotion. Kanter (1977), Larwood and Wood (1977) and Riger and Galligan (1980) offer particularly comprehensive lists of possible organization actions. They and other writers acknowledge that women start at an inherent disadvantage, and want to counter potential negative effects. Schein (1976), for example, advocates 'changes within the organization, designed to minimize the influence of stereotypical thinking on selection and promotion decisions'. Women are treated as a separate organization group who need selective help.

Many of the remedial approaches have clarity as their theme. Companies are

Table 9.1 Suggestions for increasing the number and prospects of women in management jobs

---

I: *Integrate women into current organization systems as equals*

Recruitment
Number balancing
Training
Career development
Facilitating mentor relationships and membership of informal networks

II: *Modify current practices to help women*

Flexible working hours
Child-care facilities
Part-time working
Job sharing
Returner schemes
Alternative remuneration packages

III: *Organization development with an awareness of women's issues*

IV: *Improve other organization members' acceptance of women*

V: *Train women to participate in employment in similar ways to men*

Job skills
Interpersonal skills

VI: *Advice to women on how to participate and make choices in a man's world*

How to be a manager
Career development
Combining career and having children
Join informal networks
Self responsibility

---

advised to develop clear job descriptions, clear selection criteria, clear role definitions; clear appraisal standards. In practical terms 'clarity' might also mean publicly advertising job vacancies within a company so that all employees have equal chances of applying, or examining in-company and public relations documentation to purge any overt or implicit sexism (Venables, 1981). If job requirements and skills are measured objectively, it is claimed, confusion, conflict, competition and therefore discrimination will be eliminated. Objectivity affords a way of being fair. In Schein's terms we should move from 'Think Manager—Think Male', to 'Think Manager — Think Qualified Person'.

Several writers advise organizations to tackle the issue of relative number (Chapter 4) directly, rather than expect their proportion of women in responsible jobs to increase as a result of other initiatives. The effects of their proposals that companies should aim for a 'quantum' number of women in management or include women representatives in all organization problem-solving teams may, however, be undermined by the dynamics of tokenism.

Kanter suggests recruiting women in batches and clustering them in neighbouring organizational areas (rather than spreading them thinly across the company) to take these potential moderating forces into account.

With career development particularly in mind, closer monitoring of women employees' progress, talents, and aspirations is often advocated so that these can be matched to job openings. To be effective this proposal requires considerable sensitivity and counselling skills from women managers' superiors. Many writers are concerned that traditional 'female' routes of career progression offer limited opportunities for development and often lead to dead-ends. They suggest the introduction of new jobs to act as bridges between female ghettos and other organization areas, or of whole new career paths for women designed to include the ingredients considered essential for promotion to senior management.

Even if a company is keenly interested in suggestions along the lines above, they will not find action easy. One sound first step is achieving an adequate understanding of current practices. A team from Ashridge Management Centre (1980) worked with several organizations to find out where women were in the company, and, later, to design and implement activities which would improve their career opportunities. One product of their action research approach is a diagnostic guide with which companies can carry out a comprehensive review of organization practices, particularly career ladders. It is entitled 'No barriers here: a guide to career development issues in the employment of women' (Manpower Services Commission, 1981).

A further strategy in this general category is to offer aspiring women managers formal substitutes for valuable development experiences to which men have easier, probably informal, access. They could, for example, be matched with coaches who act as mentors, or be introduced into the organization's informal social networks. The bank which participated in the Ashridge project took this area as one for attention, based on the study's findings. It set out deliberately to expose new women recruits to suitable role models and opportunities to develop potentially helpful contacts. During their induction training the newcomers met peers in the same position as themselves, successful women managers (who contributed to the courses), and senior company personnel. The organization thus created suitable conditions for valuable networks and relationships to develop. What use they then made of them was left to the individuals concerned.

Finally in this category, there is considerable concern that, as research repeatedly concludes, women receive less training than do male colleagues. In a survey of 300 Canadian women managers, for example, Vandermerwe (1979) found that just over half thought the training they received inadequate. There were also strong feelings that training offerings were too male-oriented. Viewed in this light, the common proposal that women should receive more consideration as potential course participants is not as modest as it at first sounds. Training experiences may, though, require modification to cater adequately for female participants. Some training materials, practices and themes will certainly need to be de-masculated. It may also be appropriate to run separate all-women groups when discussing areas in which their experiences are likely to differ significantly from men's. Some seminars and workshops already offer this option. I shall look in more detail at the content of training in section V.

II: MODIFY CURRENT PRACTICES

Even with broad acceptance of the status quo, it is acknowledged that some modifications to organization practices may be necessary to integrate women successfully. Most of these proposals are explicitly designed to help women have children as well as careers. Organizations are encouraged to offer some flexibility, either in day-to-day working or long-term career development, leaving individuals to do the rest by careful planning.

Venables (1981) claims that flexible work hours and child care facilities are women's two principle demands, and that given these they will sort out any remaining difficulties for themselves. Perhaps because these other two conditions are seldom met, opportunities for part-time working are also prized. They particularly affect the rate at which women are able to return to work after having children. In an accounts department able to offer part-time jobs, Cleverdon (1981) reports a 60 per cent return rate compared with a typical rate of 10 per cent. Such comparisons support Chapter 2's suggestion that practical considerations often count more significantly than motivation in determining whether a woman is in employment.

The isolated examples there are of part-time working at senior organization levels tend to be self-created by the employee, and represent unique and individual contracts between them and their employers. Miller (1980) documents six case histories to show that part-time working is possible in responsible jobs. Her examples are a senior market researcher, a personnel director, a purchasing officer, a senior consulting engineer, a television production assistant and a librarian. Their stories have three factors in common: all had already proved their competence in full-time work; they planned their work pattern to fit their employers' as well as their own needs and all interpret part-time as more flexible as well as shorter-than-normal working hours.

There are a variety of patterns through which two people can share a job. They may work half a week each, for example, or one mornings and the other afternoons. Job sharing is another area of individual initiative which appears now to be bearing fruit. The Job Sharing Project in London started in 1979, and disseminates information and advice to interested individuals, employers and trade unions. Although still in its infancy, the project had 1200 enquiries in 1980. Despite expectations that job sharing might create difficulties, especially for neatly run bureaucracies, organizations who have tried it find that both they and employees can benefit. There are indications that interest in job sh strengthening. In a bulletin documenting social changes, Zweig (1981) that four million workers — 20 per cent of the United States' 2 part-time workers — are now job sharers. It appears people are evaluating the balance between various components of their lives, and deciding that they are prepared to receive less income in return for more personal time. This trend is affecting major US companies, and consultants are being called in to advise organizations on how to design flexible working arrangements.

'Returner schemes' are one of the latest offerings from companies to female employees. Most of these are at a pilot and very tentative stage of development. People pausing in career to have a child are offered the opportunity to take a longer break than that afforded by statutory maternity leave. At a recent seminar I heard about two such schemes currently operating in the United Kingdom. One, now established for over ten years, is for doctors; the other is a pilot project for employees with management potential in a major bank. Both schemes offer individual flexibility, but require retained individuals to maintain contact through occasional work practice and possibly training sessions. The bank allows five years on this arrangement; the doctors' scheme has no top ceiling but the situation is reviewed every year and the retainee may transfer to part-time if not full-time work. Scheme organizers are interested in protecting their previous investment in the employee's training, as well as in fostering equal opportunities. Participants benefit by retaining their footing in the profession, interest in and knowledge about current developments in practice, and confidence in their own abilities.

Issues of principle and practice surround such initiatives. Some of the most hotly debated questions are whether the opportunities should be open to all employees, or only those who have previously demonstrated career potential, as seems to be happening now; whether men should be offered similar opportunities; how long to offer; how employees can best keep in touch with their profession and employer; and how to find them appropriate jobs when they return. At the moment the take-up of such schemes is likely to be small. If, however, they become standard practice they could have a significant effect on individuals' expectations and career planning.

Recognizing that women managers may have needs different from those of male colleagues, some writers have suggested that organizations should offer them different remuneration packages. A paid home help might be more appropriate to their lifestyle than a company car, or extra shopping or general personal time than free executive lunches.

III: ORGANIZATION DEVELOPMENT WITH AN AWARENESS OF WOMEN'S ISSUES

Concern about women's limited opportunities alone is often insufficient to prompt an organization into remedial action. An alternative strategy for those intent on reform is to work on a problem the organization does care about, but to develop solutions which make the most of women as resources. The Ashridge Women in Management project's methods fostered this kind of development (Ashridge Management College, 1980). In the first phase the

Responsibility for self One theme underpinning all the others is that the individual should accept responsibility for herself. She can then deal more awarely with employers, and maintain her sense of her own worth and values in relationships at work. A complementary injunction is to develop self-awareness — a realistic appreciation of her own abilities and aspirations. Learning to manage or control one's female characteristics is also widely advocated. I hope that this means creatively drawing on one's female principle rather than muting it. On a more immediately practical note, Larwood interprets self-management as women knowing their legal position, and being prepared to enforce their rights if necessary.

At one level this chapter's suggestions are sensible because they take account of the world of employment as it largely is. They will have some beneficial effects, especially those such as assertiveness training which help women develop a sense of self-worth and draw on their own distinctive characteristics. I find the overall approach of reform unsatisfactory, however, because it is grounded in limited and inherently flawed values and assumptions. It tinkers with, and seek to improve, a tired and inadequate system. In the longer term the proposed strategies are unlikely to benefit either women or society generally. And many of them will not, anyway, have their expected effects. In the next chapter I explain my cynicism.

Chapter 10

A reflective interlude

The standard proposals for improving the position of women in organizations summarized in Chapter 9 will have some beneficial impact, but their rationale is inherently flawed. The equality they offer women is the freedom to be like men, and even in this objective they are likely to prove only moderately successful. These moves for reform are essentially asymmetrical, emanating from a power base of male superior and female subordinate, and so replicate the traditional pattern of social relationships they claim to supersede. They offer only surface changes. They are also based on agentic principles, suggesting that they are largely a reformulation of the male need to control the perceived threat epitomized by women. By exploring the suggestions' various deficiencies I hope, then, to learn more about the limitations of agency unmitigated by communion.

Firstly, the proffered solutions focus on one element in a complex interacting system, and fail to appreciate its ecology. Women have been singled out as 'the problem'. This is seen particularly well in any discussion about maternity leave provisions. Its focus will tend to be on expected difficulties, and steps to minimize their impact. Creative alternatives with wide-ranging benefits, such as using temporarily vacant positions to provide development secondments for other staff, are seldom, if ever, considered. Because they have been identified as the problem, women are required to change without affecting the system as a whole. Men are not making substantial accommodations in their roles to complement women's development. At home, women are still responsible for running the house and do most of the housework. At work, many men still judge them by traditional notions of femininity, which their very participation in high-level employment violates, and reward and punish them accordingly. Women are expected to absorb the resulting lifestyle and social tensions created in the system.

Significant aspects of the wider social context, especially changing social values and patterns, were also not taken into consideration by the suggestions in Chapter 9.

Agency seeks control over opposing forces. In order to achieve even partial success, appreciations of those forces are simplified and their more complex, threatening and challenging aspects ignored. This is agentic splitting and denial. Suggestions for improving women's career development opportunities concentrate on organizations' official systems and procedures. Earlier chapters demonstrated that these are a minor part of organizational functioning. Informal networks enshrining primal valuing processes can completely undermine their operation. Official equal opportunities policies are no match for these more ingrained mechanisms, which still type male positive and female negative. Suggestions which ignore this 'fact' are doomed to eventual failure. New systems may be developed, but they will essentially perpetuate old values, perhaps with new outward appearances, unless the underlying processes by which worth is assessed are also addressed.

Legislation and official equal opportunities policies have achieved surface impacts, but it seems much prejudice has merely gone underground. Sexist attitudes are resurfacing as we face economic recession. As unemployment mounts, women's right to employment is being publicly questioned. Women are also experiencing increased practical obstacles to working. Public sector spending cuts have fallen heavily on social services, and women are being forced back reluctantly into the home to fill the gaps left in society's caring (Gibb, 1981). Both the selection of these areas of spending for pruning, and the assumption that women are wholly responsible for society's nurturing, show how little fundamental values have changed.

The standard solutions for increasing the number of women in management, then, repeat power inequalities between men and women. Women are offered equal opportunities to be like men, joining them on their terms in their world of employment. The basic devaluing of femaleness relative to maleness is not fundamentally disturbed. Because it glosses over conflicts and tensions, this offer has a limited viability. Unequal power relationships have their own self-reinforcing cycle which is only temporarily stable, and so carry with them the seeds of their own eventual destruction. Current equal opportunities policies may satisfy women's demands for a while, but are already proving insufficient.

A typical repercussion of agentic coping is that threats appear to be under control, but are actually largely unaccounted for. The limited success of equal opportunities legislation illustrates this principle. Organizations can comply with the letter of the law but evade its spirit. In 1979, Glucklich and Povall took a very gloomy view of what recent legislation had achieved. In two and a half years since the Sex Discrimination Act came into effect, only 310 individuals had taken cases to Industrial Tribunals, with a success rate of 16 per cent. A London School of Economics survey of 26 major organizations showed little activity. Most had issued equal opportunity policy statements, but their replies to other questions showed that any substantial change was unlikely. Twenty-two of the twenty-six said that they did or would discriminate against women in employment. Eliminating discrimination was seen as a low company priority, and one for which they felt little grassroots' pressure. 'Equal opportunities' was not an issue with which these organizations identified.

For similar reasons, all the main selling messages used to encourage organizations into equal opportunities practices have gone largely unheeded. Employers have been urged to take action because women represent a currently wasted resource; the nation needs more people with talent in management jobs; equal opportunities make for good personnel practice; and employers can avoid legal sanctions by complying. None of these exhortations has set companies alight with enthusiasm because they do not match their perceptions of importance. In fact legislation may paradoxically have done women a disservice. Disinterested employers can point to the statutes and formal policies and claim that women already have all the help they need.

A fourth characteristic of agentic strategies is that, if applied inappropriately, they can make initial problems worse. They have a tendency to overcontrol, based on the assumption that if something is good more of the same will be better. This is fallacious logic. The calls in the literature for greater clarity (in selection and personnel development procedures, for example) as a guarantee of equality, have this tone. They prompt official organization systems to become more formalized and bureaucratic. These already over-developed characteristics are part of the malaise in modern organizations and should not be encouraged further. More tightly controlled procedures are anyway unlikely to have the desired effects. They will be unable to take the strain generated by suppressed subjectivity and emotions, and people will be even more likely to resort to informal networks with their primal value systems to resolve uncertainty and get work done.

Above all, the proposals summarized in Chapter 9 want to avoid emotional upset to organizational members. Most advocate a low-key, invisible approach to integrating women into established systems. Every effort is made to portray women as minimal threats to men's interests. Again I can identify the persistance of implicit male–female inequality. Men are still the audience for any proposed activities, and what is possible is dependent on their approval and permission. The resulting emphasis on objectivity in suggested remedies has limited value if the underlying forces at work are essentially a-rational, are emotionally charged and concerned with whose subjectivity is allowable, as previous chapters suggest.

By avoiding a full appreciation of threats and complexity, agency working alone cuts itself off from its potential for creative coping. One repercussion of using a low-key approach to increasing the number of women managers, is the development of false justifying arguments. Protesting that offering women equal opportunities is to a company's clear economic advantage is an essentially inauthentic, largely apologetic, activity. Instead we should be addressing the devaluing mechanisms which make women an undesirable product which has to be 'sold' in this way.

Some threats are denied publicly, and yet are experienced and reacted to more covertly. It is absurd to claim that women are not in competition with men for employment *as it is currently organized and defined*. Failure to address this issue openly serves only to suppress individuals' feelings of anxiety and anger, and to prohibit valuable debate of why women should not be. Acknowledging and managing the feelings involved would allow a more comprehensive and useful understanding of the essential issues. It would also lead to an appreciation that identifying female as negative is a fundamental element of cultural values which we all, not just organization decision-makers, use and sustain.

Drawing together my arguments so far, I have two interrelated criticisms of the standard proposals for increasing the number of, and opportunities for, women in management jobs. I claim, firstly, that they will have little effect because they repeat, but officially deny, the sexist values of male positive, female negative they supposedly supersede, and that these basic inequalities of power will prevent them achieving their espoused objectives. Secondly, even if they did work, I would criticize their perpetuation of an organization system based on male values. Whilst some women may derive success, growth and satisfaction from following male models of management careers, the underlying pattern that male is the dominant and the desired norm goes unchallenged. Effort is directed at a one-way traffic of socializing women into the prevailing male world, with few, if any, opportunities for women to exert reciprocal influence on that world. Equal opportunities as it is thus officially offered is a sexist interpretation. Women want the right to be women *and* to be valued.

This chapter has constituted a critique of what 'equality' means. I have implicitly been arguing throughout that equality should not be contingent on sameness but should recognize and welcome differences and accord them equal social worth. To move towards valuing women *as women*, which is my definition of full equality, we need:

(1) to understand a social system in its complexity and transform it all rather than modify separate parts in isolation;
(2) to recognize and be prepared to address the divisions of interest which currently exist between women and men;
(3) to accept and cope with the uncertainty, threat and emotions involved.

One of the main reasons why we have not yet achieved this wholehearted movement is that the right and power to assign social worth has remained in the hands of men. By drawing on principles of agency to define current issues and offer solutions, they have further reinforced this power base. At a public level women have largely accepted men's lead and bowed to their valuable capacity for clarity. In private they are tentatively exploring ways to spin new meanings and develop strategies which owe more to communion. Chapters 6–8 showed women managers involved in this process. Their engagement with choice were impressive and illuminating, revealing an independent base of values which they strove to tap authentically. Their blending of agency with communion offers an alternative model for change.

The next chapter starts from this new base to generate alternative notions of the future. These go beyond the standard proposals considered in Chapter 9,

212

and take far less of the world as we know it for granted. In this sense they are more radical. But by recognizing and growing in attunement with latent and contextual forces, they are at the same time more modest, less heroic. Many are current, if tentative, practice, although we may have to look beyond silence and invisibility to appreciate their full nature.

Chapter 11

Re-vision: Alternative futures

Most of the suggestions made in Chapter 9 acknowledge and accept the patriarchal world as it is and are thus inherently unsound in their promise for women's equality. This last step in my journey is an attempt to free myself from these stale notions, and to look over the horizon. To escape from the more-of-the-same syndrome which is currently women's greatest enemy, society needs to develop visions of possible new tomorrows in which women truly have their worth and place, and not sacrifice them to today's short-term solutions. The following pages represent an initial step in this direction.

Reform's main strategy is to seek agentic control over potential problems; this chapter explores what communion has to offer as an alternative way forward. Despite its strengths and triumphs, agency alone is insufficient as a way of life, and particularly unable to deal with an uncertain future. Here I approach from communion blended with judicial, enhancing and protective agency. This balance of qualities seeks a full appreciation of possible threats such as gender, individual differences, complexity and human transience, and tries to tolerate the ambiguity and uncertainty they create. Communion also pays attention to the contextual factors agency typically ignores. Current shifts in social values have major implications for women's places in society, and are therefore considered below.

In the communion-based approach to change I develop in this chapter, I shall place less emphasis on specific *objectives* than on understanding and paying attention to the *processes* through which change can occur. In this sense I am returning to my initial questions of Who Are You? and, What Do You Want? I shall not propose One Right Way to be a Woman Manager, but map possible directions, and advocate elements in a process of development. From the various personal and research perspectives I have access to, I can illustrate some options more fully than others; but I want at the outset to declare my respect for women who travel other journeys and come to conclusions different from my own. There is plenty of room. Choice is again one of my main themes.

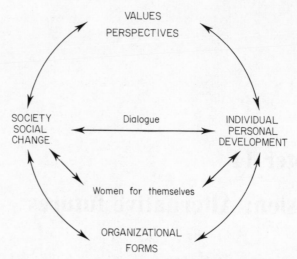

Figure 11.1   The chapter's themes in relationship

Figure 11.1 is my guide for this chapter's topics and sections. It separates out individual from society, and shows them in a dynamic, balanced tension representing the essential complementarity of the two fundamental tendencies of being: independence-seeking agency and union-seeking communion. I have called the processes which link the individual with society 'dialogue' to indicate the mutual influence which is particularly important at times of change. Personal development and social change are, at their best, mutually enhancing processes. If either occurs at the expense of the other, both will eventually be impoverished. Two major intersections of the individual with society which are relevant to women's progression in management are depicted in the figure's outer circle: values and organizational forms. When the dialogue changes, as it is doing now, these forms may be transformed.

Throughout the chapter I shall be pointing to processes of re-vision which are now affecting women, men and society, and reaching into every life area. In writing I shall, however, be more partisan. I should be disappointed to see women's current surge of energy dissipated because they cannot find appropriate ways to sustain its momentum. The 'individuals' with whom I shall therefore be most concerned are women. To pursue this bias I have included in the diagram women as their own reference group, as a subgroup of society in which many people are now finding new meaning.

SOCIETY: SHIFTING SOCIAL THEMES

I shall start by looking at society as the setting for women's current development, and review some of the social themes which seem currently to be under review, at least in the western world. The women's issues developed in this book are embedded in a context which is itself changing rapidly, and which must therefore be taken into account in attempts to achieve change. The Women's Movement and women's developing self-awareness are both fundamental elements in this broader tide of societal development, and also recipients of its impacts.

The following themes come from a variety of sources. Many have long been identified as characteristics of a post-industrial society (see Reisman, 1958; Bell, 1974), but we have so far seen few of their promised benefits. Now renewed hope is being expressed in a wave of literature which draws various strands of social development together, explores their mutual relevance and interprets them as signs of a more fundamental rising tide of new values. This literature, which claims that society has reached a crucial crossroads, has itself reached impressive proportions. From their different perspectives the following make major contributions: Gregory Bateson, 'Steps to an Ecology of Mind' (1973); Alvin Toffler, 'The Third Wave' (1980); Marilyn Ferguson, 'The Aquarian Conspiracy' (1981); and Fritjof Capra, 'The Turning Point' (1982). From them I take the following list of issues which society is now facing. Later, when I look at social values and perspectives, I shall return to these sources for the new guiding principles through which society is seeking its solutions.

Current social issues

Facing up to economic recession Unemployment; a questioning of the meaning and structure of employment and the status of unpaid work.

Changing values about work Increasing dissatisfaction amongst unskilled workers; demands for greater personal satisfaction from those with higher education levels; movement away from the Protestant work ethic.

In a volume looking ahead to the nature of work in America in the 1980s (Kerr and Rosow, 1979), Yankelovitch identifies what he calls the 'New Breed', a generation who consider themselves entitled to paid employment in 'good' jobs, but are not prepared to work particularly hard at them. In this 'psychology of entitlement', 'success' and 'self-fulfilment' have parted company. Etzioni, in the same volume, suggests we are experiencing a major shift from belief in hedonistic results to demands for hedonistic process. It is not sufficient for labour to reap eventual rewards — work must be intrinsically enjoyable. In the book's introduction, Kerr advocates 'industrialism with a human face' to meet this challenge — well-organized companies incorporating sufficient flexibility for workers to meet their individualistic needs, a society 'of *free individuals* and *controlled organizations*'.

The quality of working life movement Some individuals and organizations are deciding that stress in its various forms is no longer an acceptable price to pay for employment and productivity.

Questioning the career ethic Career progression as an end in itself is under attack from various directions. The plight of the technological specialist provides an interesting illustration. Many companies are now wondering (a) how to reward technical excellence for its own sake rather than promote capable technologists into management where they are often both unhappy and ineffective; (b) who and how to train in these areas; and (c) what to do with people whose technical skills become obsolete. Some companies, such as banks, who used to rely heavily on employee mobility are now finding that young managers are unwilling to move. Concern about the quality of other life areas seems to be moderating motivation towards a career.

Ecological concern Dwindling resources; concern about energy supplies; strengthening of anti-nuclear grassroots activity; concern about pollution in its many forms.

A suspicion of size and technology for its own sake Schumacher's 'Small is Beautiful'; appropriate technology; alternative energy sources.

Self help A movement away from professionalism; consumer associations; holistic medicine; local community action groups. A concern for personal growth—physical, emotional and spiritual. People on a rich variety of different journeys. Creating new lifestyles. A resurgence of interest in religion.

The search for community Communes; despite a chequered history, the principle of co-operative living and/or working still has many practising advocates. The search for community is often powerfully associated with complementary moves towards self-development.

Paradigm shifts Fundamental questioning of the bases of knowledge and thought. In most areas of academic enquiry — physics, biology, the social sciences— people are finding cause-and-effect patterns of thinking inadequate to appreciate fully the phenomena with which they are dealing, and are moving towards alternatives which are more capable of capturing complexity, multi-causality, discontinuous shifts in states of order, etc. Some of these developments have lead physicists to conclusions which bear more similarities to the teachings of Eastern philosophies than to traditional Western concepts about action and doing (Capra, 1976).

Women have several parts in the transformation of social values towards which the above debates or quests are building. Their energy and their questions are contributing significantly to many of its developments, particularly those which refocus attention on the private and personal as the medium for change. Women's movement into management and other high-status jobs has ironic aspects in this context. In their enthusiasm those who have accepted male norms of behaviour and goals are reinforcing the old values of organization, career and employment from which many men are retreating. It is said that whilst women are in the work revolution, men are now in the leisure revolution. Women may even be keeping alive a creaking, outmoded system, preventing its beneficial death. If this is in any way what is happening, there are sinister implications for those interested in who creates social values. If men redefine values and designate leisure or home as the valued arena just as women join in employment, women may get left behind again in devalued roles and under pressure to accept other people's definitions of the world.

A common theme to many of the above trends is a new concern to see the relationships between individual, community and ecology. In this sense they are not totally new concerns, but the surfacing of an old, enduring but muted, tradition which has always been with us in some form. This muted undercurrent complements its public sibling which is based in rationality. Essentially it is the West's own mystic tradition which has been effectively suppressed at different periods in history.

Surfacing the muted tradition

Bateson (1973) traces this muted stream back over the last two thousand years of European history. He suggests that during that period there has been a rough dichotomy, and deep controversy, between two fundamental philosophical perspectives. One position is characterized by concern about substance, the other inquires into pattern. The first asks 'What is it made of?' and the second 'How is it arranged?' Throughout history attention to substance has usually dominated, but every so often concern about pattern surfaces. Bateson identifies the Pythagoreans, Gnostics (Pagels, 1982) and Alchemists (Franz, 1980; Grossinger, 1979) as exponents of the mystic tradition. In medieval Europe witches were purged for representing similar forces (Daly, 1979).

Bateson draws on Gnostic terms to distinguish between the two fundamental philosophical traditions. He virtually takes us back to agency and communion, giving some support to Chapter 3's speculation that they are not, in fact, exact equivalents. In Gnostic philosophy the 'pleroma' and the 'creatura' represent two worlds of explanation. 'The pleroma is the world in which events are caused by forces and impacts and in which there are no "distinctions", no "differences".' This for Bateson is mind. In contrast, 'in the creatura, effects are brought about precisely by difference'. Within the realm of the creatura, agency is an appropriate *modus vivendi*. It helps us make useful distinctions which find some correspondence in the phenomena with which we deal. Applied to the pleroma, agency results in artificial splitting, the making of false distinctions. It may have its merits, in focusing one by one on parts of a car engine that is performing poorly, for example. Its dangers lie in forgetting that we are working with artificial, self-created distinctions, or that individual items make full sense only as parts of a delicately interacting whole. Our tendency to ignore wider systems in our thinking is well illustrated by environmental pollution:

'When you narrow down your epistemology and act on the premise "what interests me is me, or my organisation, or my species", you chop off consideration of other loops of the loop structure. You decide that you want to get rid of the by-products of human life and that Lake Erie will be a good place to put them. You forget that the eco-mental system called Lake Erie is part of *your* eco-mental system — and that if Lake Erie is driven insane, its insanity is incorporated in the larger system of *your* thought and experience.'
(Bateson, 1973)

One of the most significant aspects of the shifts in paradigm we are now experiencing is their common attempt to reach out through communal understanding for appropriate appreciations of the world, society and individuals in their *wholeness*. Rather than *imposing* differences, communion leaves room for sensitivity to implicit differences, and to whatever shape, including its conflicts and confusions, the world as experienced has. Sensing is not dominated by rational consciousness, but is achieved by balancing this mode with other connections we make with the world such as intuition, bodily feelings and emotions. Focusing on individual development and ignoring for the moment the constraining effects of social values, I suspect that it is easier to achieve this balance by starting in communion than by starting in agency. Whilst both modes *need* to incorporate the other to achieve individual wholeness, agency's simultaneous need for control often debars it from achieving this goal. Communion is more able to incorporate and allow a valued place for alternative strategies based in agency to make their contribution.

In practical terms, communion enhanced by agency may mean ecological awareness twinned with close observation as I address a particular issue, working with environmental forces rather than seeking only to enslave them, or embracing uncertainty without becoming swamped and disabled by anxiety. In debates between women and men about their relationship which I now often encounter, the latter seems notoriously difficult to achieve. All too often at the first expressions of powerfully felt emotions, ears close, voices are raised and positions become stereotyped and entrenched. Sometimes, though, dialogue is not prematurely aborted in this way, and a new light of possibility shines. On these occasions, it seems that a genuine intent to understand, and some measure of trust, enable both sexes to suspend immediate judgement and to experience their discomfort but continue to listen and be open to further experience.

If society is indeed surfacing a muted tradition, an important issue is where we look for guidance and direction. Established authority is steeped in the old traditions and has developed to extreme its skills of agency. It is unlikely to seek out, or be able to access, an alternative pool of values and strategies. Already new voices are finding their own forms and means of communication. The various writers about our current Turning Point urge us to identify and listen to these new sources, and to test out when their contributions are valuable.

In the 1960s parents learnt from their children. In 'Culture and Commitment', Mead (1972) points out that 'for the first time in history children had wisdom not available to their elders.' This trend to let youth have its say has continued to some extent. We are also learning that as individuals we have been restricting our own creativity by exerting excessive control on ourselves. If we look inside we find a wealth of untapped resources. Recent brain research has brought this home by distinguishing the facilities of the left and right hemispheres (Ornstein, 1975). The left (usually associated with functions on the right-hand side of the body) is concerned with order, analysis, language, time; the right with pattern, creativity, emotional tone. These two clusters of characteristics will already be very familiar! There are now moves towards re-valuing the right-brain's contribution ('intuition' *is* becoming a little more respectable), and finding ways to tap more of its potential, and of its capacity for productive cooperation with the rational left-brain. Newly popular notions about creative thinking — such as De Bono's lateral thinking (De Bono, 1970), the Inner Game technique of relaxed concentration (Gallwey, 1979) and Buzan's pattern notation (Buzan, 1974) — all operate by facilitating creative balance between the two sets of capabilities.

There is also currently a boom in learning from each other, and from ourselves through relationships with others, with a complementary shift away from revering professional experts. This swing is represented in the self-help, consciousness-raising and local community groups which are proliferating in so many diverse spheres. Miller (1976) sees such initiatives as powerful forces for social change. She advocates development via affiliation, which she feels we need to free from its old negative connotations of weakness and dependency. Growth through affiliation, she says, contains the possibility for an entirely different, more advanced, approach to living and functioning. Her suggestions provide one concrete way of putting into practice the notions of *power with* which were developed in Chapter 4.

In this search for new strategies women are centre-stage, both for the sake of their own development and with offerings for society as a whole. *They* are the main depository for the neglected, relatively untapped, muted tradition. Their suppression mirrors its suppression. Several writers note society's need for traditionally female characteristics, whether as 'attributes of the future' (Mirides and Cote, 1980), or to help men reintegrate their masculine and feminine aspects (Levinson, 1978). Women need to unveil and rediscover the muted tradition for their own sakes too if they are to become whole and valuable. I do, however, find hopeful statements that women's development will benefit society naive and simplistic. They underestimate the negative connotations of femaleness, and so ignore the hostile environment in which women's initiatives are taking place. Women are certainly not about to receive sudden welcome (from men) as potential managers on the strength of their muted characteristics. They pose too powerful a threat, as earlier chapters have shown. Women must free themselves of this context to gain the necessary space and freedom for their, and the muted tradition's, development. *They* must welcome themselves and each other. How they approach currently male dominated organizations *is* an important issue, but one which I shall temporarily put into abeyance lest it cloud my discussion of women's growth as women.

WOMEN'S DEVELOPMENT; THEIR RELATIONSHIPS WITH EACH OTHER AND WITH MEN

Both theoretically and practically, I believe that women's way forward is to develop and respect their own themes within their own culture before attempting to engage in dialogue with men. Women must become potentially self-sufficient, particularly in terms of assigning worth. Only when they have a solid base of their own can they escape from oppression without seeking to oppress in their turn. I am ambitious about what women can achieve, for themselves and others, if they do develop their own base with thoroughness and courage before attempting to change openly their social relationships with men. Various writers warn of the negative repercussions if this first phase of development is curtailed (Bardwick, 1979; Oakley, 1981). The greatest risk is that women will lose their initial energy and direction, and settle for old palliatives instead of creating new options.

The separation of women from men I advocate has indeed happened in many of the early activities of the Women's Movement. An awareness of men's potential to distort and constrain female being prompted the creation of safe spaces defended against men. This was epitomized in the development of consciousness-raising groups and of women-only magazines and newsletters. For most people this phase of self-development was a relatively private affair, perhaps fostered by belonging to a small, secret and exclusive peer group. Whilst there is some guidance available on ground-rules for running a consciousness-raising group (for example, Spender, 1980), what they do and how remain relative mysteries to outsiders and the general public. Similarly, individuals have tended to resolve issues about their gender in personalistic terms and many prefer not to share their experiences and solutions, even with people faced by similar circumstances.

I find myself both disappointed and heartened by the relative silence and invisibility of women's approaches to these issues. I am disappointed because experiences and gained understandings remain unshared. Each individual must start afresh with their own quest and may feel alone and unique because unaware of and unguided by other travellers. These trends are at the same time heartening because they represent new ways of being and becoming, free from many of the vices of doctrinaire mass movements. No competitive manifestos simplify the issues, or demand loyalty which compromises individuals' capacities for development and judgement.

The process of change through one-by-one personal transformation may seem slow, but it is sure, soundly grounded, authentic and above all continually adaptive. The critical factors for significant impact to be achieved are that numbers should grow, and that the individuals concerned should each act 'as if they make a difference' (Ferguson, 1981), thus influencing the sphere in which they operate. This is one meaning for me of the power of 'the personal is political'. In this sense, I see women as the powerful, distributed seeds of a new transformation. Their impacts challenge our stale notions that social change must be large-scale and standardized to be meaningful. They also lead me to revise my conclusion in Chapter 1 that very little is happening to improve women's places in society. I had been looking at the surface, public world and its official statistics. Turning to women's own arena, the private and personal, I find that significant transformations *are* in progress. Women's emerging form of politics provides a critical example. Women made up only 3.5 per cent of the members of Parliament in the United Kingdom in July 1983. Their representation was the same as in 1974, the year before the Sex Discrimination Act. Some commentators interpret these figures as evidence of women's continuing suppression. On a local scale, however, women are becoming increasingly active to promote causes they believe in through their own considered forms of action. Their resilient demonstration outside Greenham Common air-base, and its ripple effects promoting concordant expressions throughout the country, illustrate their alternative form of politics.

In the near future I see the possibility of taking this social transformation still further. There are signs that a new phase of collaboration between women and men is becoming possible, expanding potential life options. Friedan (1982), one of the founders of the Women's Movement in the United States, suggests that we are now moving to 'The Second Stage', of dialogue and joint development with men. The first stage of separation and opposition was necessary for women to develop their self-understanding and self-respect but is no longer the only priority.

However welcome this movement forward, women must appreciate and protect against its dangers. Women's delicately developed awareness is vulnerable in the face of the patriarchal forces which have constrained it for so long. Unless dialogue emerges out of female strength and centredness, women cannot engage on equal terms with men. Women must maintain their current channels of self-development to provide the necessary self-affirming balance. Their ability to do so is strengthened by the relatively recent emergence of positive goals within the Women's Movement, which in time can depose original formulations which defined women mainly in terms of their opposition to men. Bardwick describes this changing emphasis:

'The initial feminist analysis was preoccupied with the negative; what women had *not* become. Happily, within feminism and academia we have begun to focus on ourselves. What *have* we become? What are women's strengths, insights, and priorities? What is our history? What are our contributions? What is our tradition? Where have we found our sense of honour? . . . When we feel pride in the qualities, values, and accomplishments that have historically been ours, we will have finally given up our psychological minority status.'

One visible sign of these changes is the new tone in women's statements of their aims. Early formulations were expressed as 'we demand'. Men were seen as holding the keys to things women wanted, and their permission and approval was required before women could be free. Current rewritings of these statements more often use 'we intend', 'we assert' and 'we will' to establish principles of female action. Men, and their excess of social power, are still important, but are no longer the only reference points. In these shifts of emphasis I see the Women's Movement passing on beyond the heat of its initial anger, although this may be a necessary phase individual women still go through. Anger is most associated with seeing women's muted position from men's dominant perspective, and draws in part on men's values about what is 'good'. As women more consistently see themselves from a position which is their own but also valued, their feelings seem to become more mixed, and they recognize the choice they have about what values to adopt. They also contact more feelings of pride about the wisdom of their own female heritage.

Another precondition for satisfactory dialogue is an appropriate language of exchange. Traditional women's talk centres on experiences and feelings, men's on analysis and logic. The challenge is to find means of bridging these differences, without compromising either register or forcing either party to speak wholly in a foreign tongue. From women's viewpoint one of the main dangers of engaging in dialogue is that they will listen either too much or too little to the world around them. Listening too much, the path of communion, they will become swamped. Listening too little, they risk adopting agentic strategies, which control in their blinded need not to be controlled. In this, as in many aspects of this chapter, the strategy Ferguson (1981) terms 'Radical Centre' has much to offer. She advocates culling from both extremes of approach, and achieving an appropriate balance between, or above, them. This may mean, for example, blending assertiveness and a sense of self-worth on to a style of cooperation and showing of concern for others.

Much of the strength available to women is in their culture, and yet many facets of women-to-women communication are also in their infancy. Some women who identify with career goals, and are preoccupied with gaining acceptance in organizations as tokens, isolate themselves from other women, even (or particularly) those in similar circumstances who might form a suitable supporting network. Some who do identify themselves as 'feminist' (by which I mean willing to express an interest in and identification with women's issues) assign evaluative labels to other groups or supposed types of women. I find this a curious, agentic, wasteful use of energy (and so may, but sadly, be guilty of doing it myself). In the long term one of the most significant factors for us all will be whether women's sub-groups can tolerate, value, communicate with and learn from each other. They are in touch with or striving for alternative models of being which have the potential to go beyond stale notions of conditional relationship. They have the potential to provide new patterns for society.

As will have become plain, appropriate self-development is for me the key to successful women's dialogue, both with other women and with men. I shall now explore in more detail what this might mean for individual women.

PERSONAL DEVELOPMENT FOR WOMEN

Personal development can take various forms. Many women in management jobs have identified to a greater or lesser extent with male models and values, muting or distorting their own sense of themselves as female. I shall concentrate on exploring the implications of this starting point because I think it the more common, given education's and organizations' patriarchal base. Managers in caring jobs may have different initial priorities. They have been encouraged to overdevelop their female qualities, and may need to re-value and strengthen their own capacities for agency to achieve balance and wholeness.

A first probable step in development for women who have muted their femaleness is to appreciate fully what being a women implies, especially in a still largely male-oriented world. This core aspect of self must find its expression. If one turns to women's cultural traditions, and mythology about wise women and goddesses, it seems self-evident that woman's journey of maturation is inwards towards her own core. Whilst men quest out into the world to prove and find themselves, women carry within them their own rhythm and cycle of life and death. It is to this source of knowing they must abandon themselves to become wholly realized, both in their individual identity and in their connectedness with the world of which they are part.

From my own and other women's experiences, it seems that the first clear identification of oneself as a woman in a man's world is unavoidably disturbing and turbulent. Feeling oppressed by men as a group makes a woman angry, frustrated and upset at her previous blindness and collusion. She typically stereotypes men back in a retaliatory devaluing gesture. Relationships with individual men, particularly husbands, boyfriends and colleagues at work, become difficult. They become cultural symbols and carriers of social values against whom, however reluctantly, she directs her anger and frustration. To live through these feelings growthfully, the individual may need her own phase of separation just as the Women's Movement as a whole has done, an opportunity to withdraw into her self and discover her own strengths.

The awareness she develops during this phase of coming-to-know involves the woman changing her perspectives on the world, values, opinions and possibly lifestyle. From the upheaval of these processes a new whole emerges. She develops an appreciation of her own abilities, a clearer perception of herself in relation to others, and a sense of self-worth. Most significantly her self-acceptance is no longer contingent on male approval. It is this acceptance of self which acts as the vital base for authentic acceptance of others, the 'presence to each other' which Daly (1979) advocates. Both personal and interpersonal energy are freed in this process.

I see the dual nature of this process as particularly significant for women. By pursuing both its inward- and outward-facing aspects wholeheartedly, they can balance a strong self-affirming and self-nourishing identity with a relational identity which honours their heritage of communion. For women the choice must be both, in balance, rather than one or the other exclusively.

There is no one way of being a woman, and self-development may involve experimenting with different approaches before arriving at a profile which fits (at least for the time being). There will typically be guilt about pleasing oneself, concern about contradicting stereotypes, and anxiety about unfamiliar visibility to recognize and manage. As a society we are not used to tolerating such experiments publicly and become upset at, and therefore punish, what we see as bizarre behaviour. I think we need to allow and help people, and ourselves, to be temporarily 'crazy' as they shape new perspectives on the world. But as I expect that this will seldom happen, yet, women also continue to need models to copy, mentors and sponsors to advise them, and safe environments in which to make their more daring trials of who they are.

WOMEN-CENTRED RESOURCES FOR TRAVELLERS

For those embarking on journeys of exploration, more and more women-centred resources are opening up to facilitate and support the process. Many women are setting up local groups that will flexibly meet their needs for development, and for a community to identify with. Some are organized around focal women's issues— health, birth, motherhood or rape, for example — others have more open objectives. Most of these initiatives are well outside organizational boundaries, but are likely to benefit women members' professional competence indirectly. Within employment, women in different industries or occupations such as the media, management and computing have banded together in more visible networks. These offer formal seminars related to members' professional development, and represent women's rights within their area of employment. Awareness and debate about women's issues is also being fostered by a proliferation of seminars and local evening classes which now take women as their theme. Courses on 'Women Writers' and 'Women in Transition' are but two of the many I have come across recently.

From the feminist literature women are also being offered new images as possible models of being. There is a shift away from the early ideals which largely copied male behaviour, towards a plurality of offerings, some firmly grounded in women's traditions, others attempting to blend communion- and agency-based qualities in creative combinations. The most forceful urge is the quest for authentic anchors of womanhood. Some writers look to women's traditional social roles and seek to reinvest these with meaning. In 'Feminism: the Revised Version', Lamb (1981) reports that some founders of the Women's Movement are revising their vision of women's rights. Their general direction is away from assertive demands for whatever men have (or appear to have), and towards a renewed appreciation of women's traditional roles and of the importance of family relations. Betty Friedan, for example, is uneasy about the current emphasis on careerism, is 'considering making soup from scratch' and writes about the family as 'the nutriment of our humanness, of all our individuality'.

But as so much of women's history has been shaped by patriarchy, other writers insist that women cannot find out who they are by looking at their recent social roles. These are vestiges of women's lineage, but do not express its deeper meaning. And so a new wave of literature is leading us back to mythology, to the goddesses and gods before and during the rise of patriarchy. Colegrave (1979) looks to Chinese philosophy, and the duality and balance of yin and yang. Perera (1981) turns to the Sumerian myth of Inanna-Ishtar for her model of how modern women can seek out and come to know suppressed aspects of their identity. Inanna descends to the Underworld where she meets, submits to and is slain by her dark sister, Ereshkigal, queen of the Nether-world and of the dead. Inanna's body is placed on a peg which Perera interprets as grounding, feminine power to stand separate and potential for initiation. Ereshkigal groans and suffers at what she has done, caught up in, and embodying, the ordained connection between life and death. But descent and sacrifice bring with them the possibility of transformation. Later Inanna's release is achieved, and she returns made-anew, having touched 'the dark forces of earth reality', and achieved a new consciousness of herself in living balance.

Every so often Perera (who is a Jungian analyst) quotes sections from conversations with her female patients which resonate with and elaborate the myth's themes. She offers this assorted material as 'a way of initiation for women':

'The return to the goddess, for renewal in a feminine source-ground and spirit, is a vitally important aspect of modern woman's quest for wholeness. We women who have succeeded in the world are usually "daughters of the father" — that is, well adapted to a masculine-oriented society — and have repudiated our own full feminine instincts and energy patterns, just as the culture has maimed or derogated most of them. We need to return to and redeem what the patriarchy has often seen only as a dangerous threat and called terrible mother, dragon, or witch.'

The Inanna myth is rich in archetypes of femaleness and maleness with their different offerings. Perera takes each of its stages in turn and explores their symbolism and implications. One character, Geshtinanna, translates the themes to a mortal plane. She is a model of what women can hope to achieve by balancing their light and dark sides, and appears as the figure through whom equal and comradely relationship between woman and man eventually becomes possible.

I have suggested so far that necessary personal growth can be achieved by the individual coming to know and respect her own female grounding, and have portrayed this as an ideal path of development. But for several reasons this path may be difficult for women in management jobs to follow. However

226

reluctantly, they have accepted male norms as the price of organizational membership. As a result they may be reluctant to look inward to their own female resources or towards other women as reference points, because of the tension these encounters generate. They may well reject the more radical strands of literature which question male authority. They have already joined the traditional system this writing attacks, and have a personal investment in seeing its values survive, even though those values essentially devalue women. Several of the managers I interviewed were wary of becoming angry towards men because they thought this would make them unable to function competently and acceptably at work. The reform literature with which many managers will be more familiar compounds their estrangement from female grounding, because it uncritically takes male as its norm.

Whilst many women managers will find the stimulus and support they need to abandon the continual holding operation of muting their femaleness, others will be prompted by the relative strength of their agentic principle to take an alternative, no less potentially growthful, route. My research data showed several pushing in the opposite direction, strengthening their public workselves at the expense of their neglected, personal femaleness. Those who followed this direction wholeheartedly surprised themselves with its paradoxical outcomes. They reached a sure sense of themselves as whole people, which incorporated the personally authentic aspects of womanhood they had previously laboriously striven for.

In this chapter I have offered individual women advice very different from Chapter 9's proposals for reform. I have given priority to the question Who Are You? and expect answers to What Do You Want? to emerge, largely spontaneously, from this base. Whilst I have not advocated particular options in terms of lifestyle or identity, I have proposed a process of development in which the individual starts by coming to value herself and her individuality and then moves into dialogue with others.

ESTABLISHING NEW VALUES AND PERSPECTIVES

Creative options of being and lifestyle will emerge from the processes of individual and social development discussed above. Through dialogue new values and meta-perspectives can also emerge to act as guiding principles for new times. There are already signs that this is happening in the current wave of literature about social transformation. The identifiable directions of change seem capable of tackling fundamental existential issues against which the stereotyping and segregating of the sexes, and men's domination over women, were an inadequate defence. The major emerging themes are:

Responsibility for self Individuals are taking back power from formal organizations and experimenting with its use. They are particularly learning to make their own judgements of worth, rather than relying on external sources of valuing.

Facing uncertainty and anxiety Taking responsibility back means having to face the uncertainty which had previously been controlled by projection. People are turning to a rich range of self-development activities such as meditation, co-counselling, workshops and self-help groups to develop their emotional competence and coping abilities. These are helping them confront and manage anxiety and its major causes, such as aloneness and death. Hospices for the dying, in which openness is encouraged, are an organization form which illustrates this trend.

An appreciation of ecology: seeing wholes not parts Seeing what is, without the distorting lenses of singleminded agency, leads to an appreciation of inter-connections — within an individual's being, between individuals, and between the human species and its natural and artificial setting. Action grounded in this mode of understanding emerges as if without plan, and is appropriate to the whole, not biased by partial appreciations. At an individual level this has fostered the trend towards reconstituting artificially separated physical, emotional, cognitive and spiritual aspects of being back into one human life. Ferguson (1981) sees holistic medicine, with its emphasis on health rather than illness, as one of the most developed arms of the current social transformation.

Quality of life Many people are currently experimenting with the criteria against which they judge their lives. Money and identifiable works are giving way to more qualitative measures — job satisfaction, use of capabilities, self-actualization, overall lifestyle — which give well-being priority over material possessions. Balance rather than maximization becomes their slogan.

Tolerance, choice and adaptability As individuality, variety and complex understanding increase in social value, evaluations of appropriateness become more useful than simple judgements of 'good' and 'bad.' Similarly, notions of absolute 'truth' are being replaced by an appreciation that multiple perspectives each have their validity. A general tolerance of diversity and variety has many benefits. For example, employment can be arranged flexibly both to meet individual members' needs and facilitate a company's adaptation to the environment. Individuals need no longer conform to socially created and imposed sex roles, but can choose who to be and how to live.

The notion of choice returns me to earlier themes. Choice is an uncertain, often painful, process which many people would prefer to avoid, especially if there are no clearly agreed norms of what is acceptable. Choosing therefore requires self-responsibility and competence to deal with the anxiety it creates.

New models of how change happens Along with 'new' values, the literature of social transformation contains fresh appreciations of how change happens. Social change does not have to be gradual and incremental, or organized by authority and implemented on a large scale. Major transformations can happen suddenly as a trigger event or statement (perhaps in a novel) tips the balance towards new values and social forms for which energy has previously been building individualistically. Appreciations, and realizations in practice, of new values and new paradigms of thought are currently converging in the academic world, for example, as people from diverse fields such as physics and feminism, management and medicine find old reductionist ways of thinking inadequate and reach very similar conclusions about new directions. One of their chief conclusions is that we have neglected processes in our concern about goals and steady states, and now need to redress the balance. In this sense their models are compatible with female experience and its grounding in basic life rhythms. The managers I interviewed took a process approach to their careers. They paid attention to satisfying their immediate personal needs rather than looking ahead, and found that further opportunities tended to emerge unplanned as necessary.

For some time to come women will be the main carriers of the above themes for society, and will thus be moving against established values and coping strategies. Women aspiring to senior positions in organizations are one of the groups in the forefront of this movement. They are meeting with men in one of the more exclusive preserves of male supremacy. Drawing on the theoretical and research discussions in preceding chapters, I now have some tentative appreciation of how a woman manager can act from an authentic base of womanhood whilst operating in male-dominated organizations. In many ways she will be marching to a tune different from that enshrined in established norms of behaviour.

New values in practice: women managers

As representatives of the re-emerging muted tradition operating within established organization forms, women managers epitomise the transformation in social values and structures Western society is currently experiencing. They stand both to benefit and to suffer for their focal positioning. As a group they attract exaggerated hopes for change (exaggerated in expectations that it will be painless rather than that it can be profound), and conscious and unconscious resistance. In this changing and challenging context, individuals are faced with choices about what they attempt to achieve and who they attempt to be.

Of the potential routes forward for women managers, enhancing communion with agency offers substantial opportunities for achieving a positive identity as a woman *and* achieving a challenging and satisfying life in the largely male world of employment. Even where *men's* dominance is less pronounced, and greater balance between the principles is possible, agency remains a necessary approach to many organizational tasks, and communion alone a vulnerable and limited strategy. Combinations of the two strategies offer individuals a greater range of resources, and therefore more choice. The research material presented in Chapters 6–8 has already given many indications of what blending agency with communion means in practice. The managers' debates about their management style reflect some of the key dilemmas involved in choosing. When I met them, many could not tell whether their efforts to change to a more resilient style were successful. Much of their uncertainty concerned what criteria to judge themselves by. They oscillated between using male values that competitive behaviour is normal and female ideas of harmony. Most seemed to lack a middle ground of their own sense of appropriateness. Developing a sound base of personal values to provide this function is a core task in a woman manager's development. I shall now consider what the manager who draws on her female characteristics and culture at work faces.

The manager who acts from values I have repeatedly identified as more typically female is likely to found her management style on concern for people, and work through co-operation, rather than giving the task priority and promoting competition. Continuing with this style, finding ways of maximizing its benefits, and balancing it with 'harder' approaches as necessary to protect herself from others' opposition, will require persistence and independence. Most organization cultures require a competitive, individualistic approach from committed personnel. They will encourage the manager to conform, and withhold official and informal rewards if she does not. Several members of my research sample were faced with this dilemma. They resolved it in a variety of ways. Some persisted with their preferred approach and accepted that organizational rewards would be denied them. Others resented these consequences and were experimenting with tactics which would gain them more notice and respect from others. A third group were also moving towards what they termed 'male' tactics of managing, but for different reasons. They had been promoted despite their different approach to relationships, but found the higher organizational levels at which they were working too hostile to maintain their open, communion-based style unprotected. The challenge to women in these circumstances is how to deploy agentic strategies but hold on to their base of communion. The temptation is to copy agency as they see it in men's hands; but doing so robs women of their own grounding. Blending assertiveness and self-worth onto a style of cooperation and concern for others means, instead, remembering 'I' and 'we' simultaneously.

Another area of probable difference between the woman manager and her organization environment is in dominant patterns of thinking. When she is in touch with the qualities of her fully developed cognitive style, the female manager is likely to feel dissatisfied at the focused and detached tone of much management conversation. If she can articulate what rational, analytic thinking misses out, the manager will have moved through her dissatisfaction, and probably made a valuable contribution to the processes of discussion. Approaching from communion, women can access an understanding of the world which is particularly alive in its appreciation of interconnections, emotions and contexts. Putting this perspective across involves diverging significantly from social norms of focus and detachment. Because her contribution tends towards complexity, ambiguity and emotionality it may well be ignored, and the manager may feel (or be) personally rejected. Using agentic tactics, she can develop her ability to communicate between divergent perspectives. This requires identifying the themes, issues or dilemmas her intuition and attunement to feelings signal as important, and expressing them through an appropriate balance of clarity, ambiguity and complexity. All too often women's personal knowing gets lost in vague statements which can only be grasped by people who have the same experience. When women trust their own understanding, they are able to persist in the search for expression, and to establish dialogue by checking out what other people have heard. In potentially hostile environments, the woman manager may need deliberately to reinforce her own faith in the 'truths' she speaks from. As she faces a new challenge, for example, she can remind herself of when she last performed well by calling to mind an image of the event to promote her internal confidence, and make it more transferable to fresh situations.

The more women managers can develop their powers of knowing and speaking out, the more possible it will become for them to risk saying 'It is different for women, and this is what it is like'. Many women are reluctant to become identified with a women's viewpoint, because this too-often means serving as the projection screen for other people's stereotyped expectations. But silence serves only to reinforce these negative myths. Unless people can put a women's viewpoint unashamedly in discussions or meetings, and assert rather than argue for its validity, women as a group implicitly remain second-class citizens. Even people who establish their rights to be exceptions, never fully escape other people's lingering realizations that they are, 'after all', women. Managers need, then, to go public with their respect for and delight in the strengths they have because of being women.

Another expression of their female grounding which involves managers in potentially unwelcome visibility is acknowledging and being willing to associate with each other at work. At the moment relationships between women are discouraged by the dynamics of tokenism, and other psychological pressures for individual women to trust their survival to establishing that they are exceptions to the devalued female rules. This deliberate social isolation has multiple negative consequences. Individuals are helping to create the loneliness and isolation about which many complain. But they are cut off from more than valuable companionship, and they lose another vital strand of connection to their consciousness of themselves as women. As a result they risk being less whole as people, and denying themselves significant aspects of their creativity, effectiveness and identity. Again, established norms may be reinforced to the ultimate disadvantage of individual women. Whilst individuals remain deliberately solo, their efforts to contribute aspects of communion to the environments in which they work will be continually undermined. If, instead, managers take the risks of meeting and supporting each other, female energy can come together within organizations. Women managers and professional workers, although scarce, are now sufficiently numerous to have radical impacts on organization cultures.

When women take communal values with them into companies, I would expect two particularly profound effects on organizational functioning. An acceptance of affiliation and cooperation as a means of getting work done is the more obvious repercussion. These values would complement rather than supplant individual action and conflict which are equally necessary in their place. Secondly, women's tendency to see wholes and patterns rather than categorize makes them question false, possibly unhelpful, distinctions. By reuniting disconnected aspects of situations, different perspectives on human functioning, and focal issues with their contexts, for example, women can contribute to new perceptions which facilitate understanding and appropriate action.

If women managers choose to bring their connectedness to communion into public organizational life, they must also appreciate its vulnerabilities, and inadequacy for certain purposes. Pursuing communion as her sole orientation, the manager may be swamped and overwhelmed by other people's perspectives on the world, she may find her cooperative initiatives repelled and destroyed by others' preference for competition, and she may be unable to surface and give communicable or actionable clarity to her intricate understanding. Communion, like agency, is incomplete on its own. The challenge for the manager is to blend communion with judicial agency. In Jung's (1951) terms she must draw on her animus, her internal masculine principle, as well as live out her essential female grounding. Agency can protect communion, for example, by creating frameworks which make our approaches to uncertainty more emotionally tolerable. Agency can also clarify, focus and illuminate communion's intuitive and ecologically attuned truths, and help them towards appropriate expression. Communication augmented and extended through agency comes to a new appreciation of its own worth. Many women use agentic strategies as protective armour on the outside. Confident communion provides an alternative 'protection', a strength and wisdom tended from within.

A sensitive balancing and blending of the two modalities of being helps the individual make full use of their diverse capabilities. In Chapter 4's terms it also provides them with a wide array of power sources from which to approach their social world. For women now, personal power and the clear vision that allows them to see through social definitions and establish their own are particularly valuable. These confer flexibility, and with it choice.

As they strengthen their own base, women develop greater choice about what balance of communal grounding and agentic questing they live. They have a choice about when, and by how much, they turn their viewpoint of the world into public speech, or guard and feed off it privately. They also have choice about whether they become representatives of women's rights generally or pursue their own transformation in their own life space. They have choice about when they do and do not confront sexism and discrimination. And they have choice about how they constitute a lifestyle which fosters and develops their sense of identity. One of the most significant choices women currently face is how to participate in employment.

The majority of women would currently see themselves as having little choice about being employed in some fashion. They appreciate the opportunities for personal satisfaction paid work provides, and many need an income to support themselves, their children, and possibly their male partners. But the choice of *how* to participate in employment raises fundamental questions about a woman's sense of her own identity. Most women appreciate that organizations are male domains into which they are currently accepted on prescribed terms. Competing with men in this environment despite, and even because of, the barriers is challenging, stimulating, enjoyable and growthful. But, as many women realize only after many years, it involves giving up something of oneself as a woman. Being a woman has been so socially devalued that this requirement may well have been seen as an advantage until recently. Now, as more women reclaim and revalue their identity as women, full participation in male patterns and structures of employment requires a more significant sacrifice. Few organizations allow the necessary room and validation for women's qualities and continuing life rhythms. Artificial constraints on career development, for example, discourage them from becoming mothers, which even many of the most ambitious identify with as a part of their heritage and potential. The day-to-day pace and demands of organizational life override natural life rhythms in ways which damage health and well-being. Whilst the costs to men are only recently being fully recognized in the data on stress, women are more directly aware of harmful effects on them and respectful of their bodies.

Many women now consciously want to take who they are as women into organizations, and evaluate opportunities against this key criterion. The research material included in earlier chapters showed the kind of work environments they are looking for. Ironically some may soon be excluding themselves from organizational life despite the legal rights they have won, as they try to avoid overload and find that they simply do not fit in male-run organizations created to meet men's needs. I already see tentative moves in this direction. Some professional women have left jobs in companies to become self-employed. They say they are looking both for a less pressured lifestyle, and for working environments which fit their values and respect their full array of qualities. As I talk to final-year business studies students I find that some of the women reject jobs which they think require aggression or an over-commitment to work relative to other life areas, or have no intrinsic meaning. Others start out on traditional male routes of career, but with resolutions that if these fail, over time, to allow them to be themselves, they will look for alternatives.

Women approach employment from their own individuality and wholeness and look for what it can offer for their immediate personal development. They are continually choosing afresh as their capacities and needs change, and at each point of choice they consider a wide range of alternatives in and outside employment. Given these choices, many women *will* choose management and other professional jobs, attracted by the challenges they represent. I have already explored the conditions I feel are necessary for them to follow this route and maintain their identity as women. If they can draw on a strong self-affirming, woman-centred consciousness, they can invest the world of employment with their own meanings. This is a difficult and taxing path to follow, requiring continual awareness and balancing. It is not, however, the only option open to women who are ambitious to develop their competence. Some may prefer to start afresh rather than battle to humanize patriarchal organizations.

In the near future I see an increasing number of women appreciating the deficiencies of current employment, and searching to identify or to create types of employment which meet their needs to be women and to make full and flexible use of their choices. There are already some moves in this direction in the popularity of job sharing; but there are other opportunities, at which I can now only glimpse, which take less of current employment practice for granted. In the next section I shall consider some guidelines for new organizational forms which are either emerging from women's needs and life patterns, or compatible with their values and priorities.

NEW ORGANIZATIONAL FORMS

In Figure 11.1, which has guided this chapter's structure, I identified organizational structures and procedures as some of the major translations into practice of society's values and conflicts. The current reappraisal of social values is sufficiently fundamental and powerful to change existing organizational forms and create new ones which reflect new principles. These developments will involve women and men, but I shall concentrate here on initiatives which draw on and respect female energy. For the moment they may well exist separately, alongside their patriarchal brothers, until more joint development is appropriate.

A few writers in the women in management literature do offer more radical guidelines for reshaping organizations to make them suitable places for women. Kanter (1977), for example, advocates fluid organization structures which distribute responsibility widely, and allow variety in employee characteristics and styles of behaviour. She believes this can be achieved through relatively flat, decentralized companies in which the majority of employees are arranged in project working teams. Such structures, she says, would be intrinsically more open to all deviant and minority groups, not only women. Other popular suggestions involve replacing prescribed criteria and long-term planning with flexibility. The relaxation of normative expectations about career progression would, for example, be of obvious help to women wanting to work and have a family. Paying more attention to jobs and less to careers would benefit organizational effectiveness as well as women's opportunities. Current promotional reward systems often result in the neglect of 'work' in favour of more publicly visible aspects of 'employment' (Pym, 1980).

Freeing up employment patterns could take various forms. Individuals could have phases of no or part-time work whilst they pursued other aspects of their lives, such as being parents, travelling and taking an appropriate rest. Freer movement of employees within the organization would be necessary to provide continuous staffing, but could be used to increase the opportunities for short-term secondments, broaden individuals' experience and balance the pressures of involvement at work. In this context, maternity leave would be neither an organizational problem nor an unfair privilege granted to women, because it would make sense in a wider context of meta-values such as organizational effectiveness and personal well-being.

Another benefit of the framework I am proposing would be a movement away from reliance on age and official qualifications criteria as the most important filtering mechanisms in selection. Organizations rely on such measures to avoid making more individualistic judgements, and so often ignore talents appropriate to the job. Taking a more individual approach to recruitment and promotion would initially expose selectors to more uncertainty and anxiety than they currently handle. If, however, they *and job applicants* can develop greater personal responsibility, a more open negotiation (aware of, but not overwhelmed by, personal biases and needs) might be possible.

Some organizations may be amenable to change in these directions, but we cannot expect even good intent to achieve rapid, radical progress. An alternative path is for women to create employment spaces of their own, compatible with their own experiences and values. This would not be a move backwards into traditional female jobs, but a positive use of the vision and choice women now have available. By creating their own range of job options, women can revise employment, investing it with their own meanings.

Whilst the above suggestions were relatively piecemeal, women already have more whole potential organization models in the organizing principles which have informed and nurtured their current activities in the Women's Movement and other, local, interest groups. The current examples have promise in their vitality and authenticity to women's values, but many so far lack the delicate agency-based overlay which seems necessary to sustain them over time. The emerging prototype form is of small groups linked in networks, and possibly wider webs of networks linked in federations. Similarly organized networks are developing in certain occupational areas such as industrial health care and organizational development, and amongst those with concerns about peace, ecology and alternative technology. This organizational form has so far mainly been useful to dispersed individuals who would sometimes like the support of similarly minded people.

Membership of the small groups in this loose-knit organizational form is usually based on common interest. Typical aims are providing emotional, and possibly professional, support; sharing information and skills; and pursuing new paths of personal or occupational development as a joint venture. Often there is an implicit assumption that over time the group will develop and change, either in its membership or purposes, and that everyone will play some part in guiding its direction. For women the main initial instigations have been consciousness-raising about their identity as women, health and gynaecology, and action groups to establish women's rights, especially to control over their own bodies. Values in the small groups I am describing tend towards equality, co-operation, and tolerating and welcoming differences. Individuals offer their personal expertise, but do not use it to establish power over others. More so than in most other organizational forms, members' commitment and involvement are important, because if these wane significantly the group becomes a stale experience, or dissolves altogether. The group may also decay because its original mission has been accomplished or superseded. It will sometimes be appropriate for a group to disband altogether if this happens. Alternatively members may redefine its purpose or regroup in a new pattern.

Within the network small groups communicate through newsletters or less formal means, passing on information and pooling expertise. Typically certain individuals become focii, and they may then take on more coordinative functions. Sometimes groups across several similarly oriented networks organize a special event of common interest — a campaign to spread information about nuclear arms, for example. Temporary structures are created and dissolved as needs arise and change, and as members move or fluctuate in commitment.

Networks seem inherently fragile and temporary, and need continual inputs of energy and enthusiasm from members to maintain them. Yet they also bring their immediate intrinsic rewards of participation, and so have their own sustaining forces whilst they continue to meet individuals' needs. Their main weaknesses in practice are their lack of a supportive infrastructure and of long-term funding. They are therefore poor at surviving temporary crises of purpose or fluctuations in membership. When these happen, some groups or networks take security in elaborate bureaucracy and control, but these often destroy their original mission and vitality. As they experiment with different options, many of the networks I know are now showing that they can develop levels of coordination appropriate to sustain them *and* allow them to retain their flexibility, and can live through temporary troughs of commitment without becoming dispirited.

F International

One operating example which shows that employment can have different shapes if that is what people want is the computer consultancy F International. Its functioning illustrates many of the principles I have already identified as possibilities. I shall therefore describe the company in some depth, particularly considering how it deals with the routine tasks which have encouraged many organizations towards fixed, bureaucratic structures.

F International was set up in 1962 by Mrs 'Steve' Shirley to meet her own personal needs. Having reached a senior-programmer position she wanted to start a family, but could not find a part-time position in data processing. She established an organization whose core mission was to offer people who want to have children and a career, stimulating, responsible work on a flexible basis. F International now draws on the services of 830 skilled people in the United Kingdom, and is among the top five software houses in the country. In 1982 its turnover expanded by 30 per cent, and growth continues at about that rate. Similar operations have been set up in Denmark, Holland and the United

States, and the company is looking into the viability of operating in other EEC countries. In its day-to-day approach to its markets and use of resources, and in its longer term development as an organization, F International's hallmark is flexibility.

This flexibility is most apparent in the variety of contracts the company offers its staff. People make different contributions according to their availability and needs. Of the total available personnel, approximately 130 (15 per cent) are salaried. These make up the permanent organization structure. About half the salaried staff work full-time, and the rest are on either 15 hour or 20 hour-a-week contracts. It is sometimes decided that a particular job requires a full-time employee (posts such as Regional Manager, for example, require high levels of commitment), but this is usually not essential. So part-time workers hold very senior organization positions. The company provides careers, not just work, for its staff.

About 700 of F International's available staff are self-employed; these are the members of the technical 'panel' who work on consultancy projects. (They may work freelance for other companies at the same time.) Here the organization structure becomes even more fluid. Panel members generally have other commitments which mean they can only work if they do so from home. Most are women with children. Of the 5 per cent male staff, many are single parents or disabled, and so have similar employment needs. Panel members initially sign the company's Terms of Service, and negotiate, as work becomes available, for a particular project or time period. They then agree what work they will do and their hours of employment. They can work on several projects simultaneously if their skills are in demand *and* they want employment, or they may have little or no work for a time if they want a break or the company has no contracts to offer. Parents of school-age children can, for example, tailor their workload to meet domestic commitments.

Almost all salaried and panel personnel are home- rather than office-based, and regularly travel out from home to work. Salaried staff are loosely oriented towards a regional office, but most of their time is spent either at home or on clients' premises. Panel members work at clients' premises if not at home, except for attending regular project team meetings. To ensure availability, staff on a full-time contract agree to be 'on site' (which really means away from home) every day; those on a 20 hour contract, three days a week. But there is considerable individual freedom in how and when they meet these requirements.

Talking to F International's staff, I found their personal commitment to employment striking. All have deliberately considered whether or not to work, and retain responsibility and enthusiasm for their choice to do so. Panel members are particularly independent in this sense. They negotiate their employment terms and pattern with the regional Staffing Coordinator as they sign each new project contract, and so can regularly monitor and adjust the overall balance in their lives.

F International's structure has many characteristics of the networks described above. It is currently organized into five geographic regions. (Regionalization happened in 1978 to keep pace with growth.) The Regional Managers together with a small Head Office team — the Business Development Manager, Group Financial Controller and Group Personnel and Training Manager — make up the Senior Management Advisory Committee. This committee has considerable freedom for decision-making at its own level, and also proposes strategy to the Managing Director and the company's small non-executive Board. The role of the central committee has strengthened in recent years, as business and the wider web of the organization has grown. Now several functions, such as strategic planning and new technical projects, which could not be handled as effectively through dispersed, network working, are established at the centre before being devolved out to the regions.

The five regions are semi-autonomous profit centres. The Regional Manager works with the Managing Director on a marketing strategy for the year, and so is largely free to determine the shape of business and how the region is run. As links between the regions and Head Office, Regional Manager posts are important intersections in the organization's network. Each region was originally organized differently according to local circumstances. They have recently moved towards more common structures as the Regional Managers have had more contact with each other and with the Group Personnel and Training Manager.

The Regional Manager typically works with five functional managers covering sales, project management ('production'), staffing, technical support and finance), secretaries and administrators. All these posts, together with permanent jobs such as estimating and staffing coordination, are salaried. Many project managers are also salaried, although here the dividing line between salaried staff and the Panel becomes blurred. Some project and assistant project managers are members of the self-employed Panel. Project teams are organized around particular client needs, and vary considerably in size and timescale. Projects are part of the production function, but in practice some project managers may report to other function heads to spread the load. To keep the interface with the client simple, the project manager is their sole contact.

I was curious to find out how F International handles the routine organizational tasks of communication, coordination, maintaining standards, staffing, training, career progression and organizational cohesion. In each case I found the company twinning high professional standards with maximum flexibility.

Communications in F International are highly dependent on the telephone and postal services, and on a few key people, secretaries mainly, staying relatively still whilst the rest of the organization revolves around them. In my dealings with them, I found the company very responsive, and always quick to provide the contacts and information I needed. Regular meetings bring together project teams, regional management teams, Regional Managers, Project Managers, Sales people and Estimators in a web of regional and cross-regional contacts. These are particularly important to F International's functioning as face-to-face contacts are relatively unusual in this dispersed organization. Technical bulletins and newsletters help spread professional and personal information through the network.

At the heart of F International's technical coordination are its project management procedures. These have been developed over the years to provide a sound base for estimating and monitoring project requirements and time-scales, and for assuring quality. That the company now run training courses in project management and estimating for the Computer Industry Training Council is an indication of the industry's respect for their skills.

F International recruits technical staff who are already highly skilled in their fields. Applicants must have a minimum of four years' experience in the data processing industry, and no more than a two-year gap since last working. The company seldom advertises publicly for Panel members. Every summer a letter goes out to major data processing installations, asking them to inform any employees who are leaving to have children about F International. As most other organizations can offer part-time, low-level jobs at best to home-based workers, F International retrieves rather than competing for skilled labour (which is very scarce in the industry). The majority of applicants who are interviewed are then taken on. Only those whose skills and geographic location mean they are very unlikely to get work are advised against joining, and even they may be put on a 'pending' panel as no-one can predict what contracts will come up in the future.

The company has a mechanized staff record system, continually updated, as a database on Panel members' skills. Their training is directly project related (much of it using 'self-study' techniques to preserve the home base). Salaried staff receive more general technical and management training, and especially appreciate the contact with other members which residential courses offer. Employee development is a continuing aspect of F International's functioning, rather than a separate activity. Panel members develop new technical skills through their project work. They are also introduced to project management through the posts of Assistant Project Manager and by leading a team which includes a more experienced member.

Panel members' availability fluctuates continually as their home circum-stances change. Variations in the pool of available skills make the job of the Regional Staffing Coordinator, who matches people to projects, particularly important. Occasionally the company may be unable to take on an assignment because they do not have the necessary expertise in that area of the country at the time. Few people leave the company altogether. Even when they are practically able to manage full-time work, many prefer to maintain the balanced, multi-faceted life they have established through flexible working. Many of the salaried staff are ex-Panel members who now want more perma-nent employment. Particularly with its recent growth, the company is easily able to live up to its promise of offering staff careers. Some very senior posts have been filled from outside the company, but this is expected to become less necessary in the future.

One of prospective clients' main questions to F International is how they can ensure that project workers do a good job if they cannot constantly see and control their performance. They are wary of an organization which does not match the standard appearances of employment. F International works largely on trust. Panel members fill in time-sheets saying what they have done, and they are believed. They are highly qualified professional people, and assumed to be responsible and capable. Nonetheless the company is concerned to maintain the high standards on which it has built its reputation. Overall, fewer judgements are made about employees' performance than in other organizations; they are largely responsible for themselves. Quality assurance comes formally through project management procedures. If someone's work is unsatisfactory, this becomes apparent by the end of an assignment. Only if interviews with the Resources Manager and further work both fail to achieve an improvement, would they be asked to leave the Panel.

I have already mentioned several of the factors which make F International hold together as an organization: its communication networks, secretaries and project management procedures. As important is staff and panel members' commitment to the principles on which the organization is founded. The company offers challenging employment and career progression on the flexible terms its personnel need. By doing so it recognizes that its members not only pride themselves on high professional standards, but insist that personal aspects of their lives are also important. This twinning of technical excellence with a recognition of employees as whole people permeates the organization culture. It is acceptable to attend school sports days, for example, and to say that you are doing so. People use their flexible employment contracts to accommodate their other life interests, and do so publicly with a freedom they have not experienced in other organizations. They are also aware that few, if any, other companies would allow part-time workers into such responsible jobs. With these attitudes of commitment, it is hardly surprising that F International staff have a reputation for working very hard. Perhaps too, though, the nature of project working contributes to this image. People's activity is defined in terms of the work to be done, rather than any auxiliary characteristics of employment.

F International maximizes flexibility in its dealings with clients, and in its internal organization. It also incorporates an expectation of change over time. Its internal core aim remains the same — formally stated 'F International's policy is to utilize, wherever practicable, the services of people with dependents who are unable to work in a conventional environment'; but how to achieve this is continually under review. Now that computer technology is geared to interactive terminal working, Panel members go on site more than in the company's early days when batch processing (submitted overnight, and sometimes by post) was standard practice. Most members accept this trend, as solo homework does not satisfy all their needs for stimulation and social contact. In time, as computer equipment and communication become cheaper, the cycle may well return full circle and make home-working the norm again.

An area of major change at the moment is the company's attitude to its markets. F International are proud that 80 per cent of their turnover is from repeat business. They do, however, want to continue rapid expansion so that they can employ still more of their target workers. Besides fierce competition and markets threatened by recession, the company faces the additional obstacle that many potential clients think a remote workforce may be unmanageable. Once F International have done one project in an organization, they know they are likely to be asked back, so they are now considering whether to adopt more aggressive sales techniques to increase their new business. They are faced with similar dilemmas to the managers I interviewed, who were trying to adopt 'male tactics' to compensate the vulnerabilities of the low-key management style they preferred. So the company is in a testing phase; maintaining the 'straight' business approach they value, and emphasizing their technical abilities, but at the same time experimenting with more competitive strategies they see others use with apparent success. Like the individual managers, they are reserving judgement on whether their experiments are profitable and tally sufficiently with their image of themselves.

Looking to the future, the members of the company I spoke to expect change to continue. The regionalized, network structure seems capable of substantial growth without jeopardizing its efficiency; but they remain ready to respond as seems appropriate to changing environmental conditions.

I have offered this extensive profile of F International as a prototype for a new way of working. This organization form will not suit all work situations; but it, and similar creations on the same theme, have considerable potential. In F International's case, conventions about employment and career practices have proved much less a constraint on possibility than one might expect. The company functions in a manner similar to the networks of small groups discussed above, but has organized itself to avoid their main vulnerabilities. In its guiding principles it models a way of organizing which accommodates women's qualities and life patterns. The company is continually defining and redefining itself, taking its form by matching members' needs and potential contributions with market requirements. In this matching process, the organization structure achieved is minimal, and working procedures are more important to effective functioning than are fixed, official relationships. The company takes its shape from the employees' and clients' need which create it, rather than being a self-sustaining structure into which individuals must fit.

CONCLUDING REMARKS

This chapter has been about change, but does not end with a step-by-step manifesto for action. The important tasks at the moment are understanding and acknowledging what we face in the forms of fundamental issues, forces which hold them in place and the context within which they operate. A full communion-based appreciation of this sort brings its own impetus for change and its own directions. This book has attempted to contribute to the necessary first stage of clear vision with its various description, debates, protestations and questionings. It has also been able to identify changes which are already afoot as individuals experience transformations in their lives.

The directions of change which initially attracted me were towards re-valuing women and welcoming their social participation and contribution. I have now come to see these as parts of a wider social process of re-vision through which we all revitalize our appreciation of communal modes of being. Whilst I respect agency's strengths, it is not an appropriate strategy to cope alone with our futures. Bakan's advice to men was to mitigate agency with communion, and I remind them of the insistence in his message. I believe women's path lies in enhancing communion with agency, and that in this balanced combination they can find and express their own strengths.

Endpiece

As I finish writing, my main need is to pause and let what I have learnt settle fully into my everyday life. I am poised on a new, gentler phase in my personal journey — a phase which is more to do with being than thinking. I have passed through much of the surprise, anger and frustration expressed in this book's early chapters, and am now preparing to move on, even beyond the radical feminism I so value.

I have remembered the quiet voice of womanhood which appeared in Chapter 1 as a third perspective on women's issues, but has seldom been heard since. She leads me in several directions. One route is further inwards, towards my own self and the stories it has yet to tell. Another path leads back to ancient myths of what it is to be a woman, with their powerful symbols and paradoxes, and to other sources on the now muted tradition. Simultaneously, I am reaching out to the, still usually difficult, dialogue between women and men for which society now seems ready.

References

Allport, G.W. (1954), *The Nature of Prejudice*, Addison-Wesley.

Archer, J. (1978), 'Biological explanations of sex role stereotypes: conceptual, social and semantic issues', in *The Sex Role System: Psychological and Sociological Perspectives* (Eds. J. Chetwynd and O. Hartnett), Routledge and Kegan Paul, London.

Ashridge Management College (1980), *Employee Potential: Issues in the Development of Women*, Institute of Personnel Management, London.

Bachtold, L.M. (1976), 'Personality characteristics of women of distinction', *Psychology of Women Quarterly*, **1**, 1.

Bakan, D. (1966), *The Duality of Human Existance*, Beacon Press, Boston.

Bardwick, J.M. (1979), *In Transition*, Holt Rinehart and Winston, London.

Bartol, K.M. (1976), 'Relationship of sex and professional training area to job orientation', *J. Applied Psychology*, 61.

Bartol. K.M. (1978), 'The sex structuring of organizations: a search for possible causes', *Academy of Management Review*, October.

Bartol, K.M., and Butterfield, D.A. (1976), 'Sex effects in evaluating leaders', *J. Applied Psychology*, 61.

Bateson, G. (1973), *Steps to an Ecology of Mind*, Granada, London.

Bayes, M., and Newton, P.M. (1978), 'Women in authority: a socio-psychological analysis', *J. Applied Behavioural Science*, **14**, 1.

Beckman, G. McK. (1976), 'Legal barriers, what barriers?', *Atlanta Economic Review*, **26**, 2.

Bell, D. (1974), *The Coming of Post-Industrial Society*, Basic Books, New York.

Bentler, P.M. (1976), 'A typology of transsexualism: gender identity theory and data', *Archives of Sexual Behaviour*, **5**, 6.

Bernard, J. (1973), 'My four revolutions: an autobiographical history of the A.S.A.', in *Changing Women in a Changing Society* (Ed. J. Huber), University of Chicago Press, Chicago.

Bernard, J. (1981), *The Female World*, Free Press, New York.

Blake, R. R., and Mouton, J.S. (1966), 'Managerial facades', *Advanced Management J.*, July.

Bowman, G.W., Worthy, N.B., and Greyser, S.A. (1965), 'Are women executives people?', *Harvard Business Review*, July-August.

Bray, D.W. (1976), 'Identifying managerial talent in women', *Atlanta Economic Review*, **26**, 2.

Bray, D.W., Campbell, R.J., and Grant, D.L. (1974), *Formative Years in Business: A Long-Term AT & T Study of Managerial Lives*, Wiley, New York.

Brief, A.P., and Oliver, R.L. (1976), 'Male–female differences in work attitudes among retail sales managers', *J. Applied Psychology*, **61**, 4.

Brown, L.K. (1979), 'Women and business management: *Signs*,' *J. Women in Culture and Society*, **5**, 2.

243

244

Buzan, T. (1974), *Use Your Head*, BBC Publications, London.
Callaway, H. (1981), 'Women's perspectives: research as revision', in *Human Inquiry* (Eds. P. Reason and J. Rowan), Wiley, London.
Capra, F. (1976), *The Tao of Physics*, Fontana, London.
Capra, F. (1982), *The Turning Point: Science, Society and the Rising Culture*, Wildwood House, London.
Carlson, R. (1972), 'Understanding women: implications for personality theory and research, *J. Social Issues*, **28**, 2.
Chodorow, N. (1971), 'Being and doing: a cross-cultural examination of the socialization of males and females', in *Women in Sexist Society: Studies in Power and Powerlessness* (Eds. V. Garmick and B.K. Moran), Basic Books, New York.
Clance, P.R., and Imes, S.S. (1978), 'The impostor phenomenon in high achieving women', *Psychotherapy: Theory, Research and Practice*, **15**, 3, Fall.
Claremont de Castillejo, I. (1973), *Knowing Woman*, Harper and Row, London.
Clevedon, J. (1981), 'Editorial: Women in Management', The Business Graduate, **XI**, 1.
Colegrave, S. (1979), *The Spirit of the Valley: Androgyny and Chinese Thought*, Virago, London.
Cooper, C., and Lewis, B. (1979), 'The Femanager Boom', *Management Today*, July.
Cooperstock, R., and Lennard, H.L. (1979), 'Some social meanings of tranquiliser use', *Sociology of Health and Illness*, **I**, 3, 331–47.
Coote, A. (1978), 'In the beginning was the unburnt bra', *Sunday Times Magazine*, 1 October.
Cornillon, S.K. (1972), 'The fiction of fiction', in *Images of Women in Fiction: Feminist Perspectives* (Ed. S.K. Cornillon), Popular Press, Ohio.
Daly, M. (1979), *Gyn/Ecology*, Women's Press, London.
Dasey, R. (1981), 'Women in computing', *Women and training News*, **3**, Summer.
Davis, M. (1981), 'After all, we are only seeking to work', *Electronics Times*, 5 February.
Deaux, K. (1979), 'Sex evaluations of male and female managers', *Sex Roles*, **5**, 5.
De Bono, E. (1970), *Lateral Thinking: A Textbook of Creativity*, Ward Lock Educational, London.
Denmark, F.L. (1977), 'Styles of leadership', *Psychology of Women Quarterly*, **2**, 2.
Dickson, A. (1982), *A Woman in Your Own Right*, Quartet Books, London.
Eichenbaum, L., and Orbach, S. (1982), *Outside In . . . Inside Out*, Penguin, Harmondsworth, Middx.
Eichler, M. (1980), *The Double Standard*, Croom Helm, London.
Etzioni, A. (1979), 'Quality of life', in *Work in America* (Eds. C. Kerr and J.M. Rosow), Van Nostrand Reinhold, New York.
Farrell, W. (1975), *The Liberated Man*, Bantam.
Ferber, F., Huber, J., and Spitze, G. (1979), 'Preferences for men as bosses and professionals', *Social forces*, **58**, 2.
Ferguson, M. (1981), *The Aquarian Conspiracy*, Routledge and Kegan Paul, London.
Field, H.S., and Caldwell, B.E. (1979), 'Sex of supervisor, sex of subordinate, and subordinate job satisfaction', *Psychology of Women Quarterly*, **3**.
Fisher, C. (1978), 'Equal pay and equal opportunity', *Labour Gazette (Canada)*, **78**, 7.
Fishman, P. (1977), 'Interactional shiftwork', in *Heresies; a Feminist Publication on Arts and Politics*, No. 2, May, pp. 99–101.
Fogarty, M., Allen, A.J., Allen, I., and Walters, P. (1971), *Women in Top Jobs: Four Studies in Achievement*, Allen and Unwin, London.
Foster, L.W., and Kolinko, T. (1979), 'Choosing to be a managerial woman: an examination of individual variables and career choice', *Sex Roles*, **5**, 5.
Franz, M-L von (1980), *Alchemy: An Introduction to the Symbolism and the Psychology*, Inner City Books, Toronto.
Freire, P. (1972), *Pedagogy of the Oppressed*, Penguin, Harmondsworth.
French, J.R.P., and Raven, B. (1959), 'The bases of social power', in *Studies in Social*

Power (Ed. D. Cartwright), The Institute for Social Research, Michigan.

Friedan, B. (1982), *The Second Stage*, Michael Joseph, London.

Gallwey, W.T. (1979), *The Inner Game of Tennis*, Bantam, New York.

Garland, H. (1977), 'Sometimes nothing succeeds like success: reactions to success and failure in sex-linked occupations', *Psychology of Women Quarterly*, **2**, 1.

Gibb, F. (1981), 'Women at work: the five wasted years', *The Times*, 11 June.

Gilligan, C. (1982), *In a Different Voice: Psychological Theory and Women's Development*, Harvard University Press.

Glucklich, P., and Povall, M. (1979), 'Equal opportunities: a case for action in default of the law', *Personnel Management*, **II**, 1.

Goffman, E. (1950), *The Presentation of Self in Everyday Life*, Doubleday, Garden City, N.Y.

Gowler, D., and Legge, K. (1982), 'Dual-worker families', in *Families in Britain* (Eds. R. Rapoport and R.N. Rapoport), Routledge Kegan Paul, Henley.

Grossinger, R. (Ed) (1979), *Alchemy: pre-Egyptian Legacy, Millenial Promise*, North Atlantic Books, Richmond, California.

Hallett, L.H. (1982), 'Women in the 80s, *Options*, May.

Hammond, V. (1981), 'Men and women managers: the challenge of working together, *Women and Training News*, **5**, Winter.

Heilman, M.E., and Guzzo, R.A. (1978), 'The perceived cause of work success as a mediator of sex discrimination in organizations', *Organizational Behaviour and Human Performance*, **21**.

Heilman, M.E., and Saruwatari, L.R. (1979), 'When beauty is beastly: the effects of appearances and sex on evaluation of job applicants for managerial and non-managerial jobs', *Organizational Behaviour and Human Performance*, **23**.

Heisenberg, W. (1963), *Physics and Philosophy*, Allen and Unwin, London.

Henley, N.M. (1977), *Body Politics: Power, Sex and Nonverbal Behaviour*, Prentice Hall, Englewood Cliff, N.J..

Hennig, M., and Jardim, A. (1978), *The Managerial Woman*, Marion Boyars, London.

Horner, M.S. (1972), 'Toward an understanding of achievement — related conflicts in women', *J. Social Issues*, **28**.

Hunt, A. (1968), *A Survey of Women's Employment*, HMSO.

Inderlied, S.D., and Powell, G. (1979), 'Sex-role identity and leadership style: different labels for the same concept?', *Sex Roles*, **5**, 5.

Jung, C.G. (1951), *Collected Works, Vol. 9, Part II: Aion,* Routledge and Kegan Paul, London.

Kanter, R.M. (1977), *Men and Women of the Corporation*, Basic Books, New York.

Keller, E. (1980), 'Feminist critique of science: a forward or backward move?', *Fundamenta Scientiae*, **1**.

Kerr, C., and Rosow, J.M. (1979), *Work in America: 'The Decade Ahead'*, Van Nostrand Reinhold, London.

Korda, M. (1975), *Male Chauvinism: How it Works at Home and in the Office*, Hodder and Stoughton.

Lakoff, R. (1975), *Language and Woman's Place*, Harper and Row, New York.

Lamb, R.W. (1981), 'Feminism: the revised version', *Sunday Times*, 15 November.

Langrish, S.V. (1981), 'Why don't women progress to management jobs?', *Business Graduate*, **XI**, 1.

Larwood, L., and Blackmore, J. (1978), 'Sex discrimination in managerial selection', *Sex Roles*, **4**.

Larwood, L., and Wood, M.M. (1977), *Women in Management*, Lexington Books, Lexington, Mass.

Levinson, D.J. (1978), *The Seasons of a Man's Life*, Ballantine, New York.

Lewis, M. (1972), 'Parents and children: sex-role development', *School Review*, **80**, 229–40.

246

Little, J.B. (1981), 'Personal planning for change — reflections on recent conversations about career planning', *Business Graduate*, **XI**, 1.

Lobban, G. (1976), 'Sex-roles in reading schemes', in *Sexism in Children's Books: Facts, Figures and Guidelines*, Winters and Reades Publishing Co-operative, London.

Lukes, S. (1980), *Power: a Radical View*, Macmillan, London.

Mangham, I.L. (1978), *Interactions and Interventions in Organizations*, Wiley, London.

Manpower Services Commission (1981), *No Barriers Here: A Guide to Career Development Issues in the Employment of Women*. HMSO.

Marshall, J. (1981), 'Making sense as a personal process', in *Human Inquiry* (Eds. P. Reason and J. Rowan), Wiley, London.

Marshall, J., and Cooper, C.L. (1979), *Executives Under Pressure*, Macmillan, London.

Massengill, D., and Di Marco, N. (1979), 'Sex role stereotypes and requisite management characteristics: a current replication', *Sex Roles*, **5**, 5.

Mayes, S.S. (1979), 'Women in positions of authority: a case study of changing sex roles', *SIGNS: J. Women in Culture and Society*, **4**.

Mclane, H.J. (1980), 'Selecting, developing and retaining women executives — a corporate-strategy for the 80s', *Personnel*, 57, 5.

Mead, M. (1972), *Culture and Commitment — the New Relationships Between the Generations in the 1970s*, Doubleday, Garden City, N.Y.

Mennerick, L.A. (1975), 'Organisational structuring of sex roles in a nonstereotyped industry', *Administrative Science Quarterly*, **20**, 4, December.

Miller, J.B. (1976), *Toward a New Psychology of Women*, Beacon Press, Boston.

Miller, R. (1980), 'Eternal dilemma of working mothers', *Daily Telegraph*, 24 November.

Miller, C., and Swift, I. (1981), *The Handbook of Non-Sexist Writing*, Women's Press, London.

Miller, C., and Swift K. (1977), *Words and Women*, Gollancz, London.

Mintzberg, H. (1973), *The Nature of Managerial Work*, Harper and Row, New York.

Mirides, E., and Cote, A. (1980), 'Women in management: strategies for removing the barriers', *Personnel Administrator*, April.

Muldrow, T.W., and Bayton, J.A. (1979), 'Men and women executives and processes related to decision accuracy', *J. Applied Psychology*, **64**.

National Research Council (1980), *Women Scientists in Industry and Government: How Much Progress in the 1970's?*, Office of Publications, Washington, D.C.

Norgaard, C.T. (1980), 'Problems and perspectives of female managers' *MSU Business Topics*, **1**.

Novarra, V. (1980), *Women's Work, Men's Work*, Marion Boyars, London.

Oakley, A. (1981), *Subject Women*, Martin Robertson, Oxford.

Olsen, T. (1977), 'One out of twelve: women who are writers in our century', in *Working it Out* (Eds. S. Ruddick and P. Daniels), Pantheon, New York.

Olsen, T. (1978), *Silences*, Delacorle Press/Seymour Lawrence, New York.

Ornstein, R.E. (1975), *The Psychology of Consciousness*, Penguin Books, Harmondsworth, Middx.

Orwell, G. (1954), *Nineteen Eighty-Four*, Penguin Books, Harmondsworth, Middx.

Osborn, R.N., and Vicars, W.M. (1976), 'Sex stereotypes: an artifact in leader behavioural and subordinate satisfaction analysis', *Academy of Management* J., **19**.

Pagels, E. (1982), *The Gnostic Gospels*, Pelican, Harmondsworth, Middx.

Perera, S.B. (1981), *Descent to the Goddess: A Way of Initiation for Women*, Inner City Books, Toronto.

Petty, M.M., and Lee, K. (1975), 'Moderating effects of sex of supervisor and subordinate on relationships between supervisory behaviour and subordinate satisfaction', *J. Applied Psychology*, **60**.

Pogrebin, L.C., Van Gelder, L., and Micossi, A.L. (1978), 'Can Women Really Have it All?', *MS*, March.

Powell, G.N. (1980), 'Career-development and the women manager — a social power perspective', *Personnel*, **3**.
Pym, D. (1980), Towards the dual economy and emancipation from employment, *Futures*, **12**, 3.
Reddin, W. J. (1970), *Managerial Effectiveness*, McGraw-Hill, New York.
Reif, W.E., Newstrom, J.W., and Monczka, R.M. (1975), 'Exploding some myths about women managers', *California Management Review*, **17**, 4.
Reinharz, S. (1981), 'Experiential analysis: a contribution to feminist research methodology', in *Theories of Women's Studies Vol 2*. (Eds. G. Bowles and R. Duelli-Klein), Berkeley, California.
Reisman, D. (1958), 'Leisure and work in post industrial society', in *Mass Leisure* (Eds. E. Larrabe and R. Meyershon), Glencoe Free Press.
Renwick, P.A., and Tosi, H. (1978), 'The effects of sex, marital status, and educational background on selection decisions', *Academy of Management J.*, **21**.
Rich, A. (1979), *On Lies, Secrets and Silence: Selected Prose: 1966–78*, Norton, New York.
Richbell, S. (1976), 'De facto discrimination and how to kick the habit', *Personnel Management*, **8**, 11.
Riger, S., and Galligan, P. (1980), 'Women in management: an explanation of competing paradigms', *American Psychologist*, **35**, 10.
Roche, G.R. (1979), 'Much ado about mentors', *Harvard Business Review*, **57**.
Rosen, B., and Jerdee, T.H. (1973), 'The influence of sex-role stereotypes on evaluations of male and female supervisory behaviour', *J. Applied Psychology*, **57**.
Ruddick, S., and Daniels, P. (eds.) (1977), *Working it Out: 23 Women Writers, Artists, Scientists and Scholars Talk about their Lives and Work*, Pantheon, New York.
Sarah, E., Scott, M., and Spender, D. (1980), 'The education of feminists: the case for single-sex schools', in *Learning to Lose* (Eds. D. Spender and E. Sarah), Women's Press, London.
Schein, E.H. (1979), 'Organisational socialisation, and the profession of management', in *Organisational Psychology* (Eds. D.A. Kolb, I.M. Rubin and J.M. McIntyre), Prentice-Hall, New Jersey.
Schein, V.E. (1973). 'The Relationship between sex role stereotypes and requisite management characteristics', *J. Applied Psychology*, **57**.
Schein, V.E. (1976). 'Think manager — think male', *The Atlanta Economic Review*, March–April.
Senker, J. (1982), 'Women's jobs change focus as microelectronics step in', *Electronics Times*, 20 May.
Sharpe, S. (1975), *Just Like a Girl*, Penguin, Harmondsworth.
Showalter, E. (1977), *A Literature of their Own: British Women Novelists from Brontë to Lessing*, Princeton University Press, Princeton.
Singer, J. (1976), *Androgyny: Towards a New Theory of Sexuality*, Doubleday, Garden City, N.Y.
Spender, D. (1980), *Man made Language*, Routledge & Kegan Paul, London.
Staines, G., Tavris, C., and Jayaratne, T.E. (1974), 'The queen bee syndrome', *Psychology Today*, January, **60**.
Terman, L.M., and Tyler, L.E. (1954), 'Psychological sex differences', in *Manual of Child Psychology* (Ed. L. Carmichael), Wiley, New York.
Tkach, H. (1980), 'The female executive', *Managing*, No. 1.
Toffler, A. (1980), *The Third Wave*, Pan, London.
Tresemer, D. (1974), 'Fear of success in males and females: 1965 and 1971', *J. Consulting and Clinical Psychology*, **42**, 3.
Vandermerwe, S. (1979), 'Women as managers — current attitudes programs of Canadian businessmen', *Business Quarterly*, **44**, 1.
Venables, D. (1981), 'Paid work and the status of women', *Business Graduate*, **XI**, No. 1.

248

Vincent, S. (1978), 'Ten years of women's lib', *Sunday Times*, 1 October.

Vroom, V.H. (1973), 'Leadership', in *Handbook of Organisational Psychology* (Ed. M. Dunnette), Rand McNalty.

Waetjen, W.B., Schuerger, J.M., and Schwartz, E.B. (1979), 'Male and female managers — self concept, success and failure', *J. Psychology*, **103**, 1.

Winship, J. (1978), 'A woman's world: woman — an ideology of femininity', in *Women Take Issue*, (Women's Studies Group, University of Birmingham), Hutchinson, London.

Witkin, H.A. *et al*. (1962), *Psychological Differentiation*, Wiley, New York.

Yankelovitch, D. (1979), 'Work values, and the new breed', in *Work in America* (Eds. C. Kerr and J.M. Rosow), Van Nostrand Reinhold, New York.

Zweig, C. (1981), 'Job-sharing', *Leading Edge*, **II**, 4.

Index

250